Foundations of Discrete Mathematics

FOUNDATIONS OF DISCRETE MATHEMATICS

Ashutosh Nama

Oxford Book Company
Jaipur ▪ New Delhi

Published By
Oxford Book Company

Regd. Off. "Bony Residency", Gate No. 2
Opp. Tilak Public School, Vishweswariya Nagar,
Gopalpura Road, Jaipur - 302018, India
Ph: 0141 - 2761280, Telefax: 2761381
email - oxfordbookjpr@gmail.com
website - oxfordbookcompany.com

Delhi Office: 102, 1st Floor, Satyam House,
4327/3, Ansari Road, Darya Ganj,
New Delhi - 110002, India
Ph: 011-45652440

Distributed By
KEYA INTERNATIONAL PRESS LLP
E-mail: keyainternationalpressllp@gmail.com

First published in India, 2026

ISBN : 978-93-5524-323-2

Typeset by Shivangi Computers, Jaipur

Printed and bound in India by Thomson Press Ltd.,
New Delhi

Contents

Preface

Discrete mathematics forms the backbone of modern computing, algorithm design, cryptography, and various branches of applied mathematics. Unlike continuous mathematics, which deals with smooth and flowing quantities, discrete mathematics focuses on distinct, countable structures such as sets, graphs, logic, and combinatorics. It is a field that bridges the gap between theoretical mathematics and practical problem-solving, making it an essential discipline for students, researchers, and professionals in computer science, engineering, and beyond.

Foundations of Discrete Mathematics is designed to introduce readers to the fundamental concepts of discrete structures in a clear, logical, and systematic manner. This book caters to beginners while also serving as a comprehensive resource for those seeking to deepen their understanding of key topics. By maintaining a balance between theoretical rigor and practical applications, this guide ensures that learners not only grasp abstract principles but also recognize their real-world significance.

Each chapter is structured to build a strong conceptual foundation, covering essential topics such as logic, set theory, relations and functions, combinatorics, graph theory, and Boolean algebra. Special attention has been given to proofs and problem-solving strategies, equipping readers with analytical skills that are invaluable in both academia and industry. The inclusion of step-by-step examples, carefully curated exercises, and illustrative diagrams enhances comprehension and reinforces learning.

A distinguishing feature of this book is its emphasis on applicability. Discrete mathematics plays a crucial role in fields such as artificial intelligence, cybersecurity, network design, and software development. By integrating real-world examples and case studies, this book demonstrates how discrete mathematical principles underpin technological advancements and innovations.

Furthermore, historical insights into the evolution of key concepts provide a broader perspective on the significance of this discipline.

Foundations of Discrete Mathematics provides a comprehensive and captivating educational experience for students starting their mathematical adventure, educators wanting a structured teaching tool, and professionals wishing to hone their analytical abilities. In addition to presenting the material, this book aims to stimulate readers' interest and foster a greater understanding of the potential of discrete mathematical thinking.

— ***Ashutosh Nama***

Chapter 1

Introduction

Discrete Mathematics

Discrete mathematics is the study of mathematical structures that are fundamentally discrete rather than continuous. In contrast to real numbers that have the property of varying "smoothly", the objects studied in discrete mathematics – such as integers, graphs, and statements in logic – do not vary smoothly in this way, but have distinct, separated values.

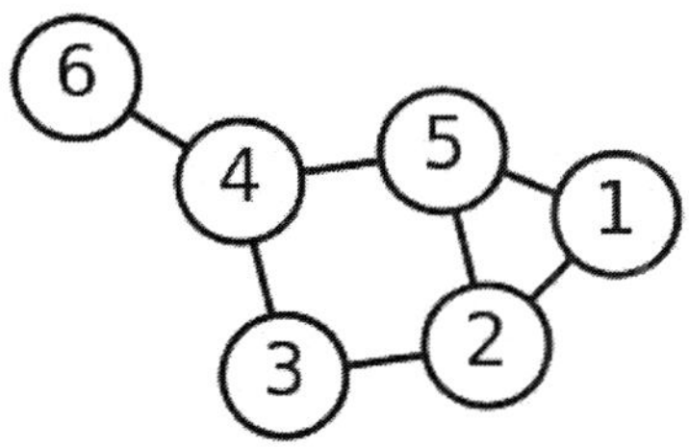

Figure: *Graphs like this are among the objects studied by discrete mathematics, for their interesting mathematical properties, their usefulness as models of real-world problems, and their importance in developing computer algorithms.*

Discrete mathematics therefore excludes topics in "continuous mathematics" such as calculus and analysis. Discrete objects can often be enumerated by integers. More formally, discrete mathematics has been characterized as the branch of mathematics dealing with countable sets (sets that have the same cardinality as subsets of the natural numbers, including rational numbers but not real numbers). However, there is no exact, universally agreed, definition of the term "discrete mathematics." Indeed, discrete mathematics is described less by what is included than by what is excluded: continuously varying quantities and related notions.

The set of objects studied in discrete mathematics can be finite or infinite. The term finite mathematics is sometimes applied to parts of the field of discrete mathematics that deals with finite sets, particularly those areas relevant to business.

Research in discrete mathematics increased in the latter half of the twentieth century partly due to the development of digital computers which operate in discrete steps and store data in discrete bits. Concepts and notations from discrete mathematics are useful in studying and describing objects and problems in branches of computer science, such as computer algorithms, programming languages, cryptography, automated theorem proving, and software development. Conversely, computer implementations are significant in applying ideas from discrete mathematics to real-world problems, such as in operations research.

Although the main objects of study in discrete mathematics are discrete objects, analytic methods from continuous mathematics are often employed as well.

Grand Challenges, Past and Present

The history of discrete mathematics has involved a number of challenging problems which have focused attention within areas of the field. In graph theory, much research was motivated by attempts to prove the four colour theorem, first stated in 1852, but not proved until 1976 (by Kenneth Appel and Wolfgang Haken, using substantial computer assistance).

In logic, the second problem on David Hilbert's list of open problems presented in 1900 was to prove that the axioms of arithmetic are consistent. Gödel's second incompleteness theorem, proved in 1931, showed that this was not possible – at least not within arithmetic itself. Hilbert's tenth problem was to determine whether a given polynomial Diophantine equation with integer coefficients has an integer solution. In 1970, Yuri Matiyasevich proved that this could not be done.

The need to break German codes in World War II led to advances in cryptography and theoretical computer science, with the first programmable digital electronic computer being developed at England's Bletchley Park. At the same time, military requirements motivated advances in operations research. The Cold War meant that cryptography remained important, with fundamental advances such as public-key cryptography being developed in the following decades. Operations research remained important as a tool in business and project management, with the critical path method being developed in the 1950s. The telecommunication industry has also motivated advances in discrete mathematics, particularly in graph theory and information theory. Formal verification of statements in logic has been necessary for software development of safety-critical systems, and advances in automated theorem proving have been driven by this need.

Computational geometry has been an important part of the computer graphics incorporated into modern video games and computer-aided design tools.

Several fields of discrete mathematics, particularly theoretical computer science, graph theory, and combinatorics, are important in addressing the challenging bioinformatics problems associated with understanding the tree of life.

Currently, one of the most famous open problems in theoretical computer science is the P = NP problem, which involves the relationship between the complexity classes P and NP. The Clay Mathematics Institute has offered a $1 million USD prize for the first correct proof, along with prizes for six other mathematical problems.

Basic Number Properties: Associative, Commutative, and Distributive

There are three basic properties of numbers, and your textbook will probably have just a little section on these properties, somewhere near the beginning of the course, and then you'll probably never see them again (until the beginning of the *next* course). My impression is that covering these properties is a holdover from the "New Math" fiasco of the 1960s. While the topic will start to become relevant in matrix algebra and calculus (and become amazingly important in advanced math, a couple years after calculus), they really don't matter a whole lot now.

Why not? Because every math system you've ever worked with has obeyed these properties! You have never dealt with a system where $a \times b$ did not in fact equal $b \times a$, for instance, or where $(a \times b) \times c$ did not equal $a \times (b \times c)$. Which is why the properties probably seem somewhat pointless to you. Don't worry about their "relevance" for now; just make sure you can keep the properties straight so you can pass the next test. The lesson below explains how I kept track of the properties.

Associative Property

In mathematics, the associative property is a property of some binary operations. In propositional logic, associativity is a valid rule of replacement for expressions in logical proofs.

Within an expression containing two or more occurrences in a row of the same associative operator, the order in which the operations are performed does not matter as long as the sequence of the operands is not changed. That is, rearranging the parentheses in such an expression will not change its value. Consider, for instance, the following equations:

$$(5+2)+1=5+(2+1)=8$$

$$5\times(5\times3)=(5\times5)\times3=75$$

Consider the first equation. Even though the parentheses were rearranged (the left side requires adding 5 and 2 first, then adding 1 to the result, whereas the right side requires adding 2 and 1 first, then 5), the value of the expression was not altered. Since this holds true when performing addition on any real numbers, we say that "addition of real numbers is an associative operation."

Associativity is not to be confused with commutativity. Commutativity justifies changing the order or sequence of the operands within an expression while associativity does not. For example,

$$(5+2)+1=5+(2+1)$$

is an example of associativity because the parentheses were changed (and consequently the order of operations during evaluation) while the operands 5, 2, and 1 appeared in exactly the same order from left to right in the expression. In contrast,

$$(5+2)+1=(2+5)+1$$

is an example of commutativity, not associativity, because the operand sequence changed when the 2 and 5 switched places.

Associative operations are abundant in mathematics; in fact, many algebraic structures (such as semigroups and categories) explicitly require their binary operations to be associative.

However, many important and interesting operations are non-associative; some examples include subtraction, exponentiation and the vector cross product.

Definition

Formally, a binary operation * on a set S is called associative if it satisfies the associative law:

$$(x * y) * z = x * (y * z) \qquad \text{for all } x, y, z \in S.$$

Here, is used to replace the symbol of the operation, which may be any symbol, and even the absence of symbol like for the multiplication.

$$(xy)z = x(yz) = xyz \qquad \text{for all } x, y, z \in S.$$

The evaluation order does not affect the value of such expressions, and it can be shown that the same holds for expressions containing *any* number of * operations. Thus, when is associative, the evaluation order can be left unspecified without causing ambiguity, by omitting the parentheses and writing simply:

$$xyz,$$

However, it is important to remember that changing the order of operations does not involve or permit moving the operands around within the expression; the sequence of operands is always unchanged.

The associative law can also be expressed in functional notation thus : $f(f(x,y),z) = f(x,f(y,z))$.

Examples

Some examples of associative operations include the following.

- The concatenation of the three strings "`hello`", "` `", "`world`" can be computed by concatenating the first two strings (giving "`hello `") and appending the third string ("`world`"), or by joining the second and third string (giving "` world`") and concatenating the first string ("hello") with the result. The two methods produce the same result; string concatenation is associative (but not commutative).

- In arithmetic, addition and multiplication of real numbers are associative; i.e.,

$$\left.\begin{aligned}(x+y)+z&=x+(y+z)=x+y+z\\(xy)z&=x(yz)=xyz\end{aligned}\right\}\text{for all } x,y,z\in\mathbb{R}.$$

Because of associativity, the grouping parentheses can be omitted without ambiguity.

- Addition and multiplication of complex numbers and quaternions is associative. Addition of octonions is also associative, but multiplication of octonions is non-associative.
- The greatest common divisor and least common multiple functions act associatively.

$$\left.\begin{aligned}\gcd(\gcd(x,y),z)&=\gcd(x,\gcd(y,z))=\gcd(x,y,z)\\\operatorname{lcm}(\operatorname{lcm}(x,y),z)&=\operatorname{lcm}(x,\operatorname{lcm}(y,z))=\operatorname{lcm}(x,y,z)\end{aligned}\right\}\text{ for all } x,y,z\in\mathbb{Z}.$$

- Taking the intersection or the union of sets:

$$\left.\begin{aligned}(A\cap B)\cap C&=A\cap(B\cap C)=A\cap B\cap C\\(A\cup B)\cup C&=A\cup(B\cup C)=A\cup B\cup C\end{aligned}\right\}\text{for all sets } A,B,C.$$

- If M is some set and S denotes the set of all functions from M to M, then the operation of functional composition on S is associative:

$$(f^\circ g)^\circ h=f^\circ(g^\circ h)=f^\circ g^\circ h \qquad \text{for all } f,g,h\in S.$$

- Slightly more generally, given four sets M, N, P and Q, with h: M to N, g: N to P, and f: P to Q, then

$$(f^\circ g)^\circ h=f^\circ(g^\circ h)=f^\circ g^\circ h$$

as before. In short, composition of maps is always associative.

- Consider a set with three elements, A, B, and C. The following operation:

×	A	B	C
A	A	A	A
B	A	B	C
C	A	A	A

is associative. Thus, for example, A(BC)=(AB)C = A. This mapping is not commutative.

- Because matrices represent linear transformation functions, with matrix multiplication representing functional composition, one can immediately conclude that matrix multiplication is associative.

Non-associativity

A binary operation $*$ on a set S that does not satisfy the associative law is called non-associative. Symbolically,

$$(x * y) * z \neq x * (y * z) \qquad \text{for some } x, y, z \in S.$$

For such an operation the order of evaluation *does* matter. For example:

- Subtraction

  $(5-3)-2 \neq 5-(3-2)$

- Division

  $(4/2)/2 \neq 4/(2/2)$

- Exponentiation

  $2^{(1^2)} \neq (2^1)^2$

Also note that infinite sums are not generally associative, for example:

$$(1-1)+(1-1)+(1-1)+(1-1)+(1-1)+(1-1)+\ldots=0$$

whereas

$$1+(-1+1)+(-1+1)+(-1+1)+(-1+1)+(-1+1)+(-1+\ldots=1$$

The study of non-associative structures arises from reasons somewhat different from the mainstream of classical algebra. One area within non-associative algebra that has grown very large is that of Lie algebras. There the associative law is replaced by the Jacobi identity. Lie algebras abstract the essential nature of infinitesimal transformations, and have become ubiquitous in mathematics. They are an example of non-associative algebras.

There are other specific types of non-associative structures that have been studied in depth. They tend to come from some specific applications. Some of these arise in combinatorial mathematics. Other examples: Quasigroup, Quasifield, Nonassociative ring.

Notation for Non-associative Operations

In general, parentheses must be used to indicate the order of evaluation if a non-associative operation appears more than once in an expression. However, mathematicians agree on a particular order of evaluation for several common non-associative operations. This is simply a notational convention to avoid parentheses.

A left-associative operation is a non-associative operation that is conventionally evaluated from left to right, i.e.,

$$\left.\begin{array}{l} x * y * z = (x * y) * z \\ w * x * y * z = ((w * x) * y) * z \\ \text{etc.} \end{array}\right\} \text{for all } w, x, y, z \in S$$

while a right-associative operation is conventionally evaluated from right to left:

$$\left.\begin{array}{l} x * y * z = x * (y * z) \\ w * x * y * z = w * (x * (y * z)) \\ \text{etc.} \end{array}\right\} \text{for all } w, x, y, z \in S$$

Both left-associative and right-associative operations occur. Left-associative operations include the following:

- Subtraction and division of real numbers:

$$x - y - z = (x - y) - z \qquad \text{for all } x, y, z \in \mathbb{R};$$

$$x / y / z = (x / y) / z \qquad \text{for all } x, y, z \in \mathbb{R} \text{ with } y \neq 0, z \neq 0.$$

- Function application:

$$(f\, x\, y) = ((f\, x) y)$$

This notation can be motivated by the currying isomorphism.

Right-associative operations include the following:

- Exponentiation of real numbers:

$$x^{y^z} = x^{(y^z)}.$$

The reason exponentiation is right-associative is that a repeated left-associative exponentiation operation would be less useful. Multiple appearances could (and would) be rewritten with multiplication:

$$(x^y)^z = x^{(yz)}.$$

- Function definition

$$\mathbb{Z} \to \mathbb{Z} \to \mathbb{Z} = \mathbb{Z} \to (\mathbb{Z} \to \mathbb{Z})$$

$$x \mapsto y \mapsto x - y = x \mapsto (y \mapsto x - y)$$

Using right-associative notation for these operations can be motivated by the Curry-Howard correspondence and by the currying isomorphism.

Non-associative operations for which no conventional evaluation order is defined include the following.

Taking the Cross product of three vectors:

$$\vec{a} \times (\vec{b} \times \vec{c}) \neq (\vec{a} \times \vec{b}) \times \vec{c} \qquad \text{for some } \vec{a}, \vec{b}, \vec{c} \in \mathbb{R}^3$$

- Taking the pairwise average of real numbers:

$$\frac{(x+y)/2+z}{2} \neq \frac{x+(y+z)/2}{2} \qquad \text{for all } x, y, z \in \mathbb{R} \text{ with } x \neq z.$$

- Taking the relative complement of sets $(A \setminus B) \setminus C$ is not the same as $A \setminus (B \setminus C)$. (Compare material nonimplication in logic.)

Distributive Property

In abstract algebra and logic, distributivity is a property of binary operations that generalizes the distributive law from elementary algebra. In propositional logic, distribution refers to two valid rules of replacement. The rules allow one to reformulate conjunctions and disjunctions within logical proofs.

For example, in arithmetic:

2 × (1 + 3) = (2 × 1) + (2 × 3) but 2 /(1 + 3) is not equal to (2 / 1) + (2 / 3).

In the left-hand side of the first equation, the 2 multiplies the sum of 1 and 3; on the right-hand side, it multiplies the 1 and the 3 individually, with the products added afterwards. Because these give the same final answer (8), we say that multiplication by 2 *distributes* over addition of 1 and 3. Since we could have put any real numbers in place of 2, 1, and 3 above, and still have obtained a true equation, we say that multiplication of real numbers *distributes* over addition of real numbers.

Definition

Given a set S and two binary operators $\cdot$ and $+$ on S, we say that the operation $\cdot$

- is *left-distributive* over + if, given any elements x, y, and z of $S, x \cdot (y + z) = (x \cdot y) + (x \cdot z)$;
- is *right-distributive* over + if, given any elements x, y, and z of S:

  $(y + z) \cdot x = (y \cdot x) + (z \cdot x)$;
- is *distributive* over + if it is left- and right-distributive.

Notice that when $\cdot$ is commutative, then the three above conditions are logically equivalent.

Propositional logic

Rule of replacement: In standard truth-functional propositional logic, *distribution* are two valid rule of replacement. The rules allow one to distribute certain logical connectives within logical expressions in logical proofs. The rules are:

$$(P \wedge (Q \vee R)) \Leftrightarrow ((P \wedge Q) \vee (P \wedge R))$$

and

$$(P \vee (Q \wedge R)) \Leftrightarrow ((P \vee Q) \wedge (P \vee R))$$

where "$\Leftrightarrow$" is a metalogical symbol representing "can be replaced in a proof with."

Truth Functional Connectives

Distributivity is a property of some logical connectives of truth-functional propositional logic. The following logical equivalences demonstrate that distributivity is a property of particular connectives. The following are truth-functional tautologies.

Distribution of conjunction over conjunction

$$(P \wedge (Q \wedge R)) \leftrightarrow ((P \wedge Q) \wedge (P \wedge R))$$

Distribution of conjunction over disjunction

$$(P \wedge (Q \vee R)) \leftrightarrow ((P \wedge Q) \vee (P \wedge R))$$

Distribution of disjunction over conjunction

$$(P \vee (Q \wedge R)) \leftrightarrow ((P \vee Q) \wedge (P \vee R))$$

Distribution of disjunction over disjunction

$$(P \vee (Q \vee R)) \leftrightarrow ((P \vee Q) \vee (P \vee R))$$

Distribution of implication

$$(P \rightarrow (Q \rightarrow R)) \leftrightarrow ((P \rightarrow Q) \rightarrow (P \rightarrow R))$$

Distribution of implication over equivalence

$$P \rightarrow (Q \leftrightarrow R) \leftrightarrow ((P \rightarrow Q) \leftrightarrow (P \rightarrow R))$$

Distribution of disjunction over equivalence

$$(P \vee (Q \leftrightarrow R)) \leftrightarrow ((P \vee Q) \leftrightarrow (P \vee R))$$

Double distribution

$$((P \wedge Q) \vee (R \wedge S)) \leftrightarrow (((P \vee R) \wedge (P \vee S)) \wedge ((Q \vee R) \wedge (Q \vee S)))$$

$$((P \vee Q) \wedge (R \vee S)) \leftrightarrow (((P \wedge R) \vee (P \wedge S)) \vee ((Q \wedge R) \vee (Q \wedge S)))$$

Examples

1. Multiplication of numbers is distributive over addition of numbers, for a broad class of different kinds of numbers ranging from natural numbers to complex numbers and cardinal numbers.
2. Multiplication of ordinal numbers, in contrast, is only left-distributive, not right-distributive.
3. The cross product is left- and right-distributive over vector addition, though not commutative.
4. Matrix multiplication is distributive over matrix addition, though also not commutative.
5. The union of sets is distributive over intersection, and intersection is distributive over union.
6. Logical disjunction ("or") is distributive over logical conjunction ("and"), and conjunction is distributive over disjunction.
7. For real numbers (and for any totally ordered set), the maximum operation is distributive over the minimum operation, and vice-versa: $\max(a,\min(b,c)) = \min(\max(a,b),\max(a,c))$ and $\min(a,\max(b,c)) = \max(\min(a,b),\min(a,c))$.
8. For integers, the greatest common divisor is distributive over the least common multiple, and vice-versa:

gcd(a,lcm(b,c)) = lcm(gcd(a,b),gcd(a,c)) and lcm(a,gcd(b,c)) = gcd(lcm(a,b),lcm(a,c)).

9. For real numbers, addition distributes over the maximum operation, and also over the minimum operation: $a + \max(b,c) = \max(a+b,a+c)$ and $a + \min(b,c) = \min(a+b,a+c)$.

Distributivity and Rounding

In practice, the distributive property of multiplication (and division) over addition may appear to be compromised or lost because of the limitations of arithmetic precision. For example, the identity $^1/_3 + {}^1/_3 + {}^1/_3 = (1+1+1)/3$ appears to fail if the addition is conducted in decimal arithmetic; however, if many significant digits are used, the calculation will result in a closer approximation to the correct results.

For example, if the arithmetical calculation takes the form: $0.33333+0.33333+0.33333 = 0.99999 \neq 1$, this result is a closer approximation than if fewer significant digits had been used. Even when fractional numbers can be represented exactly in arithmetical form, errors will be introduced if those arithmetical values are rounded or truncated.

For example, buying two books, each priced at £14.99 before a tax of 17.5%, in two separate transactions will actually save £0.01, over buying them together: £14.99×1.175 = £17.61 to the nearest £0.01, giving a total expenditure of £35.22, but £29.98×1.175 = £35.23. Methods such as banker's rounding may help in some cases, as may increasing the precision used, but ultimately some calculation errors are inevitable.

Distributivity in Rings

Distributivity is most commonly found in rings and distributive lattices. A ring has two binary operations (commonly called "+" and "*"), and one of the requirements of a ring is that * must distribute over +. Most kinds of numbers (example 1) and matrices (example 4) form rings. A lattice is another kind of algebraic structure with two binary operations, $\wedge$ and $\vee$. If either of these operations (say $\wedge$) distributes over the other ($\vee$), then $\vee$ must also distribute over $\wedge$, and the lattice is called distributive.

Examples 4 and 5 are Boolean algebras, which can be interpreted either as a special kind of ring (a Boolean ring) or a special kind of distributive lattice (a Boolean lattice). Each interpretation is responsible for different distributive laws in the Boolean algebra. Examples 6 and 7 are distributive lattices which are not Boolean algebras.

Failure of one of the two distributive laws brings about near-rings and near-fields instead of rings and division rings respectively. The operations are usually configured to have the near-ring or near-field distributive on the right but not on the left.

Rings and distributive lattices are both special kinds of rigs, certain generalizations of rings. Those numbers in example 1 that don't form rings at least form rigs. Near-rigs are a further generalization of rigs that are left-distributive but not right-distributive; example 2 is a near-rig.

Generalizations of Distributivity

In several mathematical areas, generalized distributivity laws are considered. This may involve the weakening of the above conditions or the extension to infinitary operations. Especially in order theory one finds numerous important variants of distributivity, some of which include infinitary operations, such as the infinite distributive law; others being defined in the presence of only *one* binary operation, such as the according definitions and their relations are given in the article distributivity (order theory). This also includes the notion of a completely distributive lattice.

In the presence of an ordering relation, one can also weaken the above equalities by replacing = by either ≤ or ≥. Naturally, this will lead to meaningful concepts only in some situations. An application of this principle is the notion of sub-distributivity as explained in the article on interval arithmetic.

In category theory, if (S, μ, η) and (S', μ', η') are monads on a category C, a distributive law $S.S' \to S'.S$ is a natural transformation $\lambda : S.S' \to S'.S$ such that (S', λ) is a lax map of monads $S \to S$ and (S, λ) is a colax map of monads $S' \to S'$. This is exactly the data needed to define a monad structure on $S'.S$: the multiplication map is $S'\mu.\mu'S^2.S'\lambda S$ and the unit map is η'$S.\eta$. A generalized distributive law has also been proposed in the area of information theory.

Commutative Property

In mathematics, a binary operation is commutative if changing the order of the operands does not change the result. It is a fundamental property of many binary operations, and many mathematical proofs depend on it. The commutativity of simple operations, such as multiplication and addition of numbers, was for many years implicitly assumed and the property was not named until the 19th century when mathematics started to become formalized. By contrast, division and subtraction are *not* commutative.

Common Uses

The *commutative property* (or *commutative law*) is a property associated with binary operations and functions. Similarly, if the commutative property holds for a pair of elements under a certain binary operation then it is said that the two elements *commute* under that operation.

Set Theory

In group and set theory, many algebraic structures are called commutative when certain operands satisfy the commutative property. In higher branches of mathematics, such as analysis and linear algebra the commutativity of well known operations (such as addition and multiplication on real and complex numbers) is often used (or implicitly assumed) in proofs.

Mathematical Definitions

The term "commutative" is used in several related senses.

1. A binary operation $*$ on a set S is called *commutative* if:

   $x * y = y * x \qquad \text{for all } x, y \in S$

   An operation that does not satisfy the above property is called noncommutative.

2. One says that x *commutes* with y under $*$ if:

   $$x * y = y * x$$

3. A binary function $f : A \times A \to B$ is called *commutative* if:

   $f(x,y) = f(y,x) \qquad \text{for all } x, y \in A$

History and Etymology

Records of the implicit use of the commutative property go back to ancient times. The Egyptians used the commutative property of multiplication to simplify computing products. Euclid is known to have assumed the commutative property of multiplication in his book *Elements*. Formal uses of the commutative property arose in the late 18th and early 19th centuries, when mathematicians began to work on a theory of functions. Today the commutative property is a well known and basic property used in most branches of mathematics.

The first recorded use of the term *commutative* was in a memoir by François Servois in 1814, which used the word *commutatives* when describing functions that have what is now called the commutative property. The word is a combination of the French word *commuter* meaning "to substitute or switch" and the suffix *-ative* meaning "tending to" so the word literally means "tending to substitute or switch." The term then appeared in English in *Philosophical Transactions of the Royal Society* in 1844.

Examples

Commutative operations in everyday life:

- Putting on socks resembles a commutative operation, since which sock is put on first is unimportant. Either way, the result (having both socks on), is the same.
- The commutativity of addition is observed when paying for an item with cash. Regardless of the order the bills are handed over in, they always give the same total.

Commutative operations in mathematics:

Two well-known examples of commutative binary operations:

- The addition of real numbers is commutative, since

$$y + z = z + y \qquad \text{for all } y, z \in \mathbb{R}$$

For example 4 + 5 = 5 + 4, since both expressions equal 9.

- The multiplication of real numbers is commutative, since

$$yz = zy \qquad \text{for all } y, z \in \mathbb{R}$$

For example, $3 \times 5 = 5 \times 3$, since both expressions equal 15.

- Some truth functions are also commutative, since the truth tables for the functions are the same when one changes the order of the operands.

For example, E*V*pq*V*qp; EA*pq*A*qp*; ED*pq*D*qp*; EE*pq*E*qp*; EJ*pq*J*qp*; EK*pq*K*qp*; EX*pq*X*qp*; EO*pq*O*qp*.

- Further examples of commutative binary operations include addition and multiplication of complex numbers, addition and scalar multiplication of vectors, and intersection and union of sets.

Noncommutative operations in everyday life:

- Concatenation, the act of joining character strings together, is a noncommutative operation. For example

  $EA + T = EAT \neq TEA = T + EA$

- Washing and drying clothes resembles a noncommutative operation; washing and then drying produces a markedly different result to drying and then washing.
- Rotating a book 90° around a vertical axis then 90° around a horizontal axis produces a different orientation than when the rotations are performed in the opposite order.
- The twists of the Rubik's Cube are noncommutative. This can be studied using group theory.

Noncommutative operations in mathematics:

Some noncommutative binary operations:

- Subtraction is noncommutative, since $0-1 \neq 1-0$
- Division is noncommutative, since $1/2 \neq 2/1$
- Some truth functions are noncommutative, since the truth tables for the functions are different when one changes the order of the operands.

For example, EB*pq*C*qp*; EC*pq*B*qp*; EF*pq*G*qp*; EG*pq*F*qp*; EH*pq*I*qp*; EI*pq*H*qp*; EL*pq*M*qp*; EM*pq*L*qp*.

- Matrix multiplication is noncommutative since

$$\begin{bmatrix} 0 & 2 \\ 0 & 1 \end{bmatrix} = \begin{bmatrix} 1 & 1 \\ 0 & 1 \end{bmatrix} \cdot \begin{bmatrix} 0 & 1 \\ 0 & 1 \end{bmatrix} \neq \begin{bmatrix} 0 & 1 \\ 0 & 1 \end{bmatrix} \cdot \begin{bmatrix} 1 & 1 \\ 0 & 1 \end{bmatrix} = \begin{bmatrix} 0 & 1 \\ 0 & 1 \end{bmatrix}$$

- The vector product (or cross product) of two vectors in three dimensions is anti-commutative, i.e., b × a = –(a × b).

Mathematical Structures and Commutativity

- A commutative semigroup is a set endowed with a total, associative and commutative operation.
- If the operation additionally has an identity element, we have a commutative monoid
- An abelian group, or *commutative group* is a group whose group operation is commutative.
- A commutative ring is a ring whose multiplication is commutative. (Addition in a ring is always commutative.)
- In a field both addition and multiplication are commutative.

Non-commuting Operators in Quantum Mechanics

In quantum mechanics as formulated by Schrödinger, physical variables are represented by linear operators such as x (meaning multiply by x), and d/dx. These two operators do not commute as may be seen by considering the effect of their products $x(d/dx)$ and $(d/dx)x$ on a one-dimensional wave function $\psi(x)$:

$$x\frac{d}{dx}\psi = x\psi' \neq \frac{d}{dx}x\psi = \psi + x\psi'$$

According to the uncertainty principle of Heisenberg, if the two operators representing a pair of variables do not commute, then that pair of variables are mutually complementary, which means they cannot be simultaneously measured or known precisely. For example, the position and the linear momentum of a particle are represented respectively (in the x-direction) by the operators x and $(\hbar/i)d/dx$ (where $\hbar$ is the reduced Planck constant). This is the same example except for the constant $(\hbar/i)$, so again the operators do not commute and the physical meaning is that the position and linear momentum in a given direction are complementary.

Properties of Zero

Zero is sometimes called nought (naught) or nothing or nil or O (pronounced oh). Zero is a very strange number. It is neither positive or negative. If you add or subtract zero to any number, that number stays the same. If you multiply any number by zero, you get zero. Any number raised to the power of zero is one, so $2^0=1$. However, it gets even worse. You cannot divide a number by zero. You cannot take the zeroth root of a number. It's hard to say what 0^0 is!

If you have a zero remainder in a division, then you have a whole number as an answer.

In Boolean logic, zero represents False.

There is no year zero in the BC / AD year numbering system. When the numbering goes from BC (before Christ) to AD (anno domini) the year numbering goes: 3BC, 2BC, 1BC, 1AD, 2AD, 3AD.

BC is sometimes known known as BCE (before common era) and AD as CE (common era).

If you want to count how many things there are, for example, how many apples there are in a bowl, then you start counting from one, or you end up with the wrong answer. But computer people and mathematicians sometimes start counting from zero. Fibonacci numbers start from zero, so does the series for evaluating e.

History of Zero

Zero has two meanings. It can be a number in its own right. "I have two apples, and give you two. I now have zero apples." This idea seems normal to us, but it confused the ancient Greeks, and medieval people. How can you say that *nothing* is a number? If you have one or two apples, then fine - you can see them, and see that their number is one of their properties. You can owe someone an apple (because they gave you one and you ate it, so now you must find another apple to give back) so you have have minus one apple.

But no apples is nothing. You might just as well say that you have no aardvarks! However, the Indian mathematician Brahmagupta understood that zero was a real number, and gave rules for its use in 628 AD, which were mostly accurate (he ran into trouble with dividing by zero). Once Arabic numbers were accepted, the meaning of the number zero became understood. It would be impossible to have modern mathematics without it.

The other meaning of zero is as a place holder. Early number systems such as the Egyptians had no zero. They didn't need one as they had a unary system. Two was two of the symbols for one, and three was three of those symbols, and so on. By ten, the number of symbols were getting out of hand, so they introduced a new symbol meaning ten, which also got repeated for twenty, thirty and so on. There were also symbols for hundred, thousand and so on. They tended to have a particular order, but they didn't need to. You could muddle the digits up, and still work out how big the number was. The classical Greeks had a different system, with different letters for each digit from one to nine, and yet more digits for ten, twenty and so on. But they didn't need a zero either. The Babylonians developed a positional system. This used the position of the numbers to show how big they were. We have a positional system as well, so 123 is bigger than 97, although the digits 9 and 7 are bigger than 1, 2 or 3. You put numbers in columns, and there is a units column, a tens column, a hundreds column, and so on. This means that there may be no digit in one particular column, such as two hundreds, no tens and five units. We would write this as 205 and not think twice about it. The middle zero is a place holder - "this column is empty". But the Babylonians took a long time to realise that they needed a place holder. They left a space, but spaces have a way getting left out. Anyway, what about a number like 2001? Do you leave a bigger space? Eventually, the Babylonians developed a place holder, but only within a number, not at the end. It was rather a clumsy digit . Later on, by 130 AD, Ptolemy the Greek astronomer was using the Babylonian number system, but with zero represented by a circle.

It was the Indians again who developed a consistant use of the zero. It started as a dot. Fibonacci helped to introduce the Indian number system (now called Arabic numbers) into Europe. He said "The nine Indian figures are: 9 8 7 6 5 4 3 2 1. With these nine figures, and with the sign 0 ... any number may be written." It sounds as if he didn't realise that zero was a digit!

Interestingly enough, the Mayans developed a positional number system using zero. Obviously enough, this didn't influence any European number system.

Paradox

I said above that you cannot divide by zero. In Mathematics, you need to watch out for this. It can lead to odd results! Look carefully at the argument below.

Let	a	=	b
Multiply both sides by a:	a^2	=	ab
Add $a^2 - 2ab$:	$a^2 + a^2 - 2ab$	=	$ab + a^2 - 2ab$
Simplify:	$2(a^2 - ab)$	=	$a^2 - ab$
Divide by $a^2 - ab$:	2	=	1

What went wrong? It's the line which tells you to divide by $a^2 - ab$. Since $a = b$, you have divided by zero, and this leads to the final incorrect statement.

Equality (Mathematics)

In mathematics equality is a relationship between two quantities or, more generally two mathematical expressions, asserting that the quantities have the same value or that the expressions represent the same mathematical object. The equality between A and B is written $A = B$.

x = y

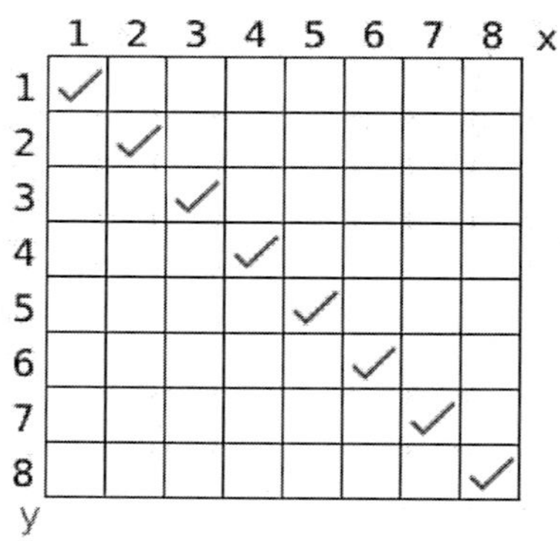

When A and B may be viewed as functions of some variables, then $A = B$ means that A and B define the same function. Such an equality of functions is called an identity. For example $(x + 1)^2 = x^2 + 2x + 1$.

In mathematical logic, equality is formalized with axioms (e.g. the first few Peano axioms, or the axiom of extensionality in ZF set theory). It can also be viewed as a relation: the identity relation, or diagonal relation, the binary relation on a set X defined by

$$\{(x,x) \mid x \in X\}.$$

The identity relation is the archetype of the more general concept of an equivalence relation on a set: those binary relations which are reflexive, symmetric, and transitive. The identity relation is an equivalence relation. Conversely, let R be is an equivalence relation, and let us denote by x^R the equivalence class of x, consisting of all elements z such that $x\ R\ z$. Then the relation $x\ R\ y$ is equivalent with the equality $x^R = y^R$. It follows that equality is the smallest equivalence relation on any set S, in the sense that it is the relation that has the smallest equivalence classes (every class is reduced to a single element).

One must not confuse equality and equation, although they are written similarly. An equality is an assertion, while an equation is the problem of finding values of some variables, called *unknowns*, to get an equality. *Equation* may also refer to an equality relation that is satisfied only for the values of the variables that one is interested on. For example $x^2 + y^2 = 1$ is the *equation* of the unit circle.

The etymology of the word is from the Latin aequalis, meaning uniform or identical, from aequus, meaning "level, even, or just."

Logical Formulations

Equality is always defined such that things that are equal have all and only the same properties. Some people define equality as congruence. Often equality is just defined as identity.

A stronger sense of equality is obtained if some form of Leibniz's law is added as an axiom; the assertion of this axiom rules out "bare particulars"—things that have all and only the same properties but are not equal to each other—which are possible in some logical formalisms. The axiom states that two things are equal if they have all and only the same properties.

Formally: Given any x and y, $x = y$ if, given any predicate P, $P(x)$ if and only if $P(y)$.

In this law, the connective "if and only if" can be weakened to "if"; the modified law is equivalent to the original.

Instead of considering Leibniz's law as an axiom, it can also be taken as the *definition* of equality. The property of being an equivalence relation, as well as the properties given below, can then be proved: they become theorems. If a=b, then a can replace b and b can replace a.

Some Basic Logical Properties of Equality

The substitution property states:

- For any quantities a and b and any expression $F(x)$, if $a = b$, then $F(a) = F(b)$ (if either side makes sense, i.e. is well-formed).

In first-order logic, this is a schema, since we can't quantify over expressions like F (which would be a functional predicate).

Some specific examples of this are:

- For any real numbers a, b, and c, if $a = b$, then $a + c = b + c$ (here $F(x)$ is $x + c$);
- For any real numbers a, b, and c, if $a = b$, then $a - c = b - c$ (here $F(x)$ is $x - c$);
- For any real numbers a, b, and c, if $a = b$, then $ac = bc$ (here $F(x)$ is xc);
- For any real numbers a, b, and c, if $a = b$ and c is not zero, then $a/c = b/c$ (here $F(x)$ is x/c).

The reflexive property states:

For any quantity a, $a = a$.

This property is generally used in mathematical proofs as an intermediate step.

The symmetric property states:

- For any quantities a and b, if $a = b$, then $b = a$.

The transitive property states:

- For any quantities a, b, and c, if $a = b$ and $b = c$, then $a = c$.

The binary relation "is approximately equal" between real numbers or other things, even if more precisely defined, is not transitive (it may seem so at first sight, but many small differences can add up to something big). However, equality almost everywhere *is* transitive.

Although the symmetric and transitive properties are often seen as fundamental, they can be proved, if the substitution and reflexive properties are assumed instead.

Relation with Equivalence and Isomorphism

In some contexts, equality is sharply distinguished from *equivalence* or *isomorphism*. For example, one may distinguish *fractions* from *rational numbers,* the latter being equivalence classes of fractions: the fractions 1/2 and 2/4 are distinct as fractions, as different strings of symbols, but they "represent" the same rational number, the same point on a number line. This distinction gives rise to the notion of a quotient set.

Similarly, the sets

$$\{A,B,C\} \text{ and } \{1,2,3\}$$

are not equal sets – the first consists of letters, while the second consists of numbers – but they are both sets of three elements, and thus isomorphic, meaning that there is a bijection between them, for example

$$A \mapsto 1, B \mapsto 2, C \mapsto 3.$$

However, there are other choices of isomorphism, such as

$$A \mapsto 3, B \mapsto 2, C \mapsto 1,$$

and these sets cannot be identified without making such a choice – any statement that identifies them "depends on choice of identification". This distinction, between equality and isomorphism, is of fundamental importance in category theory, and is one motivation for the development of category theory.

Set (Abstract Data Type)

In computer science, a set is an abstract data structure that can store certain values, without any particular order, and no repeated values. It is a computer implementation of the mathematical concept of a finite set. Unlike most other collection types, rather than retrieving a specific element from a set, one typically tests a value for membership in a set. Some set data structures are designed for static or frozen sets that do not change after they are constructed. Static sets allow only query operations on their elements — such as checking whether a given value is in the set, or enumerating the values in some arbitrary order. Other variants, called dynamic or mutable sets, allow also the insertion and deletion of elements from the set.

An abstract data structure is a collection, or aggregate, of data. The data may be booleans, numbers, characters, or other data structures. If one considers the structure yielded by packaging or indexing, there are four basic data structures:

1. unpackaged, unindexed: bunch
2. packaged, unindexed: set
3. unpackaged, indexed: string (sequence)
4. packaged, indexed: list (array)

In this view, the contents of a set are a bunch, and isolated data items are elementary bunches (elements). Whereas sets *contain* elements, bunches *consist of* elements.

Further structuring may be achieved by considering the multiplicity of elements (sets become multisets, bunches become hyperbunches) or their homogeneity (a record is a set of fields, not necessarily all of the same type).

Implementations

A set can be implemented in many ways. For example, one can use a list, ignoring the order of the elements and taking care to avoid repeated values. Sets are often implemented using various flavors of trees, tries, or hash tables.

A set can be seen, and implemented, as a (partial) associative array, in which the value of each key-value pair has the unit type.

Type Theory

In type theory, sets are generally identified with their indicator function: accordingly, a set of values of type A may be denoted by 2^A or $\mathcal{P}(A)$. (Subtypes and subsets may be modelled by refinement types, and quotient sets may be replaced by setoids.) The characteristic function F of a set S is defined as:

$$F(x) = \begin{cases} 1, & \text{if } x \in S \\ 0, & \text{if } x \notin S \end{cases}$$

In theory, many other abstract data structures can be viewed as set structures with additional operations and/or additional axioms imposed on the standard operations. For example, an abstract heap can be viewed as a set structure with a min(*S*) operation that returns the element of smallest value.

Operations

Core set-theoretical Operations: One may define the operations of the algebra of sets:

- *union(S,T):* returns the union of sets *S* and *T*.
- *intersection(S,T):* returns the intersection of sets *S* and *T*.
- *difference(S,T):* returns the difference of sets *S* and *T*.
- *subset(S,T)*: a predicate that tests whether the set *S* is a subset of set *T*.

Static Sets

Typical operations that may be provided by a static set structure *S* are:

- *is_element_of(x,S):* checks whether the value *x* is in the set *S*.
- *is_empty(S):* checks whether the set *S* is empty.
- *size(S) or cardinality(S):* returns the number of elements in *S*.
- *iterate(S):* returns a function that returns one more value of *S* at each call, in some arbitrary order.
- *enumerate(S):* returns a list containing the elements of *S* in some arbitrary order.
- *build*($x_1, x_2, \ldots, x_n$,)*:* creates a set structure with values $x_1, x_2, \ldots, x_n$.
- *create_from(collection):* creates a new set structure containing all the elements of the given collection or all the elements returned by the given iterator.

Dynamic Sets

Dynamic set structures typically add:

- *create():* creates a new, initially empty set structure.
  - * *create_with_capacity(n):* creates a new set structure, initially empty but capable of holding up to n elements.
- *add(S,x):* adds the element x to S, if it is not present already.
- *remove(S, x):* removes the element x from S, if it is present.
- *capacity(S):* returns the maximum number of values that S can hold.

Some set structures may allow only some of these operations. The cost of each operation will depend on the implementation, and possibly also on the particular values stored in the set, and the order in which they are inserted.

Additional Operations

There are many other operations that can (in principle) be defined in terms of the above, such as:

- *pop(S):* returns an arbitrary element of S, deleting it from S.
- *map(F,S):* returns the set of distinct values resulting from applying function F to each element of S.
- *filter(P,S):* returns the subset containing all elements of S that satisfy a given predicate P.
- *fold(A_0,F,S):* returns the value $A_{|S|}$ after applying $A_{i+1} := F(A_i, e)$ for each element e of S.
- *clear(S):* delete all elements of S.
- *equal(S_1, S_2):* checks whether the two given sets are equal (i.e. contain all and only the same elements).
- *hash(S):* returns a hash value for the static set S such that if equal(S_1, S_2) then hash(S_1) = hash(S_2)

Other operations can be defined for sets with elements of a special type:

- *sum(S):* returns the sum of all elements of *S* for some definition of "sum". For example, over integers or reals, it may be defined as fold(0, add, *S*).
- *nearest(S,x):* returns the element of *S* that is closest in value to *x* (by some metric).

Implementations

Sets can be implemented using various data structures, which provide different time and space trade-offs for various operations. Some implementations are designed to improve the efficiency of very specialized operations, such as nearest or union. Implementations described as "general use" typically strive to optimize the element_of, add, and delete operations.

As sets can be interpreted as a kind of map (by the indicator function), sets are commonly implemented in the same way as maps (associative arrays), namely, a self-balancing binary search tree for sorted sets (which has O(log n) for most operations), or a hash table for unsorted sets (which has O(1) average-case, but O(n) worst-case, for most operations). A sorted linear hash table may be used to provide deterministically ordered sets.

Other popular methods include arrays. In particular a subset of the integers 1..*n* can be implemented efficiently as an *n*-bit bit array, which also support very efficient union and intersection operations. A Bloom map implements a set probabilistically, using a very compact representation but risking a small chance of false positives on queries.

The Boolean set operations can be implemented in terms of more elementary operations (pop, clear, and add), but specialized algorithms may yield lower asymptotic time bounds. If sets are implemented as sorted lists, for example, the naive algorithm for union(*S*,*T*) will take code proportional to the length *m* of *S* times the length *n* of *T*; whereas a variant of the list merging algorithm will do the job in time proportional to *m+n*. Moreover, there are specialized set data structures (such as the union-find data structure) that are optimized for one or more of these operations, at the expense of others.

Language Support

One of the earliest languages to support sets was Pascal; many languages now include it, whether in the core language or in a standard library.

- Java offers the Set interface to support sets (with the HashSet class implementing it using a hash table), and the SortedSet sub-interface to support sorted sets (with the TreeSet class implementing it using a binary search tree).
- Apple's Foundation framework (part of Cocoa) provides the Objective-C classes NSSet, NSMutableSet, NSCountedSet, NSOrderedSet, and NSMutable OrderedSet. The CoreFoundation APIs provide the CFSet and CFMutableSet types for use in C.
- Python has built-in set and frozenset types since 2.4, and since Python 3.0 and 2.7, supports non-empty set literals using a curly-bracket syntax, e.g.: {x, y, z}.
- The .NET Framework provides the generic HashSet and SortedSet classes that implement the generic ISet interface.
- Smalltalk's class library includes Set and IdentitySet, using equality and identity for inclusion test respectively. Many dialects provide variations for compressed storage (NumberSet, CharacterSet), for ordering (OrderedSet, SortedSet, etc.) or for weak references (WeakIdentitySet).
- Ruby's standard library includes a set module which contains Set and SortedSet classes that implement sets using hash tables, the latter allowing iteration in sorted order.
- OCaml's standard library contains a Set module, which implements a functional set data structure using binary search trees.
- The GHC implementation of Haskell provides a Data.Set module, which implements a functional set data structure using binary search trees.
- The Tcl Tcllib package provides a set module which implements a set data structure based upon TCL lists.

As noted in the previous section, in languages which do not directly support sets but do support associative arrays, sets can be emulated using associative arrays, by using the elements as keys, and using a dummy value as the values, which are ignored.

In C++

In C++, the Standard Template Library (STL) provides the set template class, which implements a sorted set using a binary search tree; SGI's STL also provides the hash_set template class, which implements a set using a hash table.

In sets, the elements themselves are the keys, in contrast to sequenced containers, where elements are accessed using their (relative or absolute) position. Set elements must have a strict weak ordering.

Some of the member functions in C++ and their description is given in the table below:

set member functions	
Signature(s)	**Description**
`iterator begin();`	Returns an iterator to the first element of the set.
`iterator end();`	Returns an iterator just before the end of the set.
`bool empty() const;`	Checks if the set container is empty (i.e. has a size of 0)
`iterator find(const key_type &x) const;`	Searches the container for an element x and if found, returns an iterator to it, or else returns an iterator to set::end

`void insert ( Input Iterator first, Input Iterator last );`
 	
`pair<iterator, bool> insert ( const value_type& a );`
`iterator insert ( position(iterator), const value_type& a );`
	Inserts an element into the set. The first version returns pair, with its member pair::first set to an iterator pointing to either the newly inserted element or to the element that already had its same value in the set. The pair::second element in the pair is set to true if a new element was inserted or false if an element with the same value existed. And the second version returns an iterator either pointing to newly inserted element or to element having same value in set.

`void clear();`	Removes all elements in the set, making its size 0.

Multiset

A generalization of the notion of a set is that of a multiset or bag, which is similar to a set but allows repeated ("equal") values (duplicates). This is used in two distinct senses: either equal values are considered *identical,* and are simply counted, or equal values are considered *equivalent,* and are stored as distinct items. For example, given a list of people (by name) and ages (in years), one could construct a multiset of ages, which simply counts the number of people of a given age.

Alternatively, one can construct a multiset of people, where two people are considered equivalent if their ages are the same (but may be different people and have different names), in which case each pair (name, age) must be stored, and selecting on a given age gives all the people of a given age.

Formally, it is possible for objects in computer science to be considered "equal" under some equivalence relation but still distinct under another relation. Some types of multiset implementations will store distinct equal objects as separate items in the data structure; while others will collapse it down to one version (the first one encountered) and keep a positive integer count of the multiplicity of the element.

As with sets, multisets can naturally be implemented using hash table or trees, which yield different performance characteristics.

The set of all bags over type T is given by the expression bag T. If by multiset one considers equal items identical and simply counts them, then a multiset can be interpreted as a function from the input domain to the non-negative integers (natural numbers), generalizing the identification of a set with its indicator function. In some cases a multiset in this counting sense may be generalized to allow negative values, as in Python.

- C++'s Standard Template Library implements both sorted and unsorted multisets. It provides the multiset class for the sorted multiset, as a kind of associative container, which implements this multiset using a self-balancing binary search tree. It provides the unordered_multiset class for the unsorted multiset, as a kind of unordered associative containers, which implements this multiset using a hash table. The unsorted multiset is standard as of C++11; previously SGI's STL provides the hash_multiset class, which was copied and eventually standardized.
- For Java, third-party libraries provide multiset functionality:
    - Apache Commons Collections provides the Bag and SortedBag interfaces, with implementing classes like HashBag and TreeBag.

  - Google Collections provides the Multiset interface, with implementing classes like HashMultiset and TreeMultiset.
- Apple provides the NSCountedSet class as part of Cocoa, and the CFBag and CFMutableBag types as part of CoreFoundation.
- Python's standard library includes collections.Counter, which is similar to a multiset.
- Smalltalk includes the Bag class, which can be instantiated to use either identity or equality as predicate for inclusion test.

Where a multiset data structure is not available, a workaround is to use a regular set, but override the equality predicate of its items to always return "not equal" on distinct objects (however, such will still not be able to store multiple occurrences of the same object) or use an associative array mapping the values to their integer multiplicities (this will not be able to distinguish between equal elements at all).

Typical operations on bags:

- contains(B, x): checks whether the element x is present (at least once) in the bag B
- is_sub_bag(B_1, B_2): checks whether each element in the bag B_1 occurs in B_1 no more often than it occurs in the bag B_2; sometimes denoted as $B_1 \wedge B_2$.
- count(B, x): returns the number of times that the element x occurs in the bag B; sometimes denoted as $B \# x$.
- scaled_by(B, n): given a natural number n, returns a bag which contains the same elements as the bag B, except that every element that occurs m times in B occurs $n * m$ times in the resulting bag; sometimes denoted as $n \otimes B$.
- union(B_1, B_2): returns a bag that containing just those values that occur in either the bag B_1 or the bag B_2, except that the number of times a value x occurs in the resulting bag is equal to $(B_1 \# x) + (B_2 \# x)$; sometimes denoted as $B_1 \cup B_2$.

Multisets in SQL

In relational databases, a table can be a (mathematical) set or a multiset, depending on the presence on unicity constraints on some columns (which turns it into a candidate key).

SQL allows the selection of rows from a relational table: this operation will in general yield a multiset, unless the keyword DISTINCT is used to force the rows to be all different, or the selection includes the primary (or a candidate) key.

In ANSI SQL the MULTISET keyword can be used to transform a subquery in a collection expression:

SELECT expression1, expression2... FROM table_name... is a general select that can be used as *subquery expression* of another more general query, while MULTISET(SELECT expression1, expression2... FROM table_name...) transforms the subquery into a *collection expression* that can be used in another query, or in assignment to a column of appropriate collection type.

Chapter 2

Set Theory

Set theory is the branch of mathematics that studies sets, which are collections of objects. Although any type of object can be collected into a set, set theory is applied most often to objects that are relevant to mathematics. The language of set theory can be used in the definitions of nearly all mathematical objects.

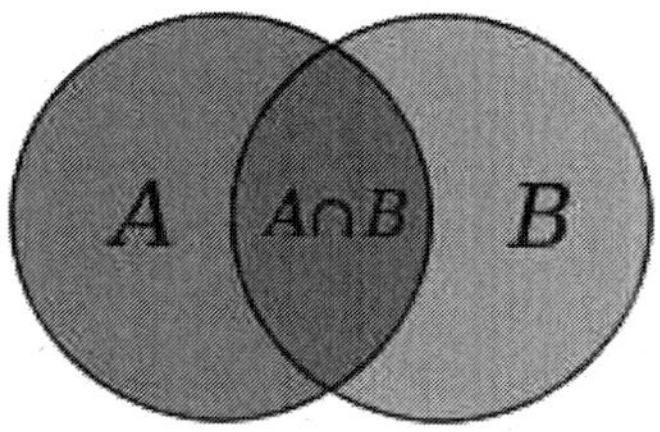

Figure: *A Venn diagram illustrating the intersection of two sets.*

The modern study of set theory was initiated by Georg Cantor and Richard Dedekind in the 1870s. After the discovery of paradoxes in naive set theory, numerous axiom systems were proposed in the early twentieth century, of which the Zermelo–Fraenkel axioms, with the axiom of choice, are the best-known.

Set theory is commonly employed as a foundational system for mathematics, particularly in the form of Zermelo–Fraenkel set theory with the axiom of choice. Beyond its foundational role, set theory is a branch of mathematics in its own right, with an active research community. Contemporary research into set theory includes a diverse collection of topics, ranging from the structure of the real number line to the study of the consistency of large cardinals.

History

Mathematical topics typically emerge and evolve through interactions among many researchers. Set theory, however, was founded by a single paper in 1874 by Georg Cantor: "On a Characteristic Property of All Real Algebraic Numbers".

Since the 5th century BC, beginning with Greek mathematician Zeno of Elea in the West and early Indian mathematicians in the East, mathematicians had struggled with the concept of infinity. Especially notable is the work of Bernard Bolzano in the first half of the 19th century. Modern understanding of infinity began in 1867–71, with Cantor's work on number theory. An 1872 meeting between Cantor and Richard Dedekind influenced Cantor's thinking and culminated in Cantor's 1874 paper.

Cantor's work initially polarized the mathematicians of his day. While Karl Weierstrass and Dedekind supported Cantor, Leopold Kronecker, now seen as a founder of mathematical constructivism, did not. Cantorian set theory eventually became widespread, due to the utility of Cantorian concepts, such as one-to-one correspondence among sets, his proof that there are more real numbers than integers, and the "infinity of infinities" ("Cantor's paradise") resulting from the power set operation. This utility of set theory led to the article "Mengenlehre" contributed in 1898 by Arthur Schoenflies to Klein's encyclopedia.

The next wave of excitement in set theory came around 1900, when it was discovered that Cantorian set theory gave rise to several contradictions, called antinomies or paradoxes. Bertrand Russell and Ernst Zermelo independently found the simplest and best known paradox, now called Russell's paradox: consider "the set of all sets that are not members of themselves", which leads to a contradiction since it must be a member of itself, and not a member of itself. In 1899 Cantor had himself posed the question "What is the cardinal number of the set of all sets?", and obtained a related paradox. Russell used his paradox as a theme in his 1903 review of continental mathematics in his *The Principles of Mathematics*.

The momentum of set theory was such that debate on the paradoxes did not lead to its abandonment. The work of Zermelo in 1908 and Abraham Fraenkel in 1922 resulted in the set of axioms ZFC, which became the most commonly used set of axioms for set theory. The work of analysts such as Henri Lebesgue demonstrated the great mathematical utility of set theory, which has since become woven into the fabric of modern mathematics. Set theory is commonly used as a foundational system, although in some areas category theory is thought to be a preferred foundation.

Set (Mathematics)

In mathematics, a set is a collection of distinct objects, considered as an object in its own right. For example, the numbers 2, 4, and 6 are distinct objects when considered separately, but when they are considered collectively they form a single set of size three, written {2,4,6}. Sets are one of the most fundamental concepts in mathematics. Developed at the end of the 19th century, set theory is now a ubiquitous part of mathematics, and can be used as a foundation from which nearly all of mathematics can be derived. In mathematics education, elementary topics such as Venn diagrams are taught at a young age, while more advanced concepts are taught as part of a university degree.

Definition

A set is a well defined collection of objects. The objects that make up a set (also known as the elements or members of a set) can be anything: numbers, people, letters of the alphabet, other sets, and so on. Georg Cantor, the founder of set theory, gave the following definition of a set at the beginning of his *Beiträge zur Begründung der transfiniten Mengenlehre*:

A set is a gathering together into a whole of definite, distinct objects of our perception [Anschauung] and of our thought – which are called elements of the set.

Sets are conventionally denoted with capital letters. Sets A and B are equal if and only if they have precisely the same elements.

As discussed below, the definition given above turned out to be inadequate for formal mathematics; instead, the notion of a "set" is taken as an undefined primitive in axiomatic set theory, and its properties are defined by the Zermelo–Fraenkel axioms. The most basic properties are that a set "has" elements, and that two sets are equal (one and the same) if and only if every element of one is an element of the other.

Describing Sets

There are two ways of describing, or specifying the members of, a set. One way is by intensional definition, using a rule or semantic description:

A is the set whose members are the first four positive integers.

B is the set of colours of the French flag.

The second way is by extension – that is, listing each member of the set. An extensional definition is denoted by enclosing the list of members in curly brackets:

$$C = \{4, 2, 1, 3\}$$

$$D = \{\text{blue, white, red}\}.$$

Every element of a set must be unique; no two members may be identical. (A multiset is a generalized concept of a set that relaxes this criterion.) All set operations preserve this property. The order in which the elements of a set or multiset are listed is irrelevant (unlike for a sequence or tuple). Combining these two ideas into an example

$$\{6, 11\} = \{11, 6\} = \{11, 6, 6, 11\}$$

because the extensional specification means merely that each of the elements listed is a member of the set.

For sets with many elements, the enumeration of members can be abbreviated. For instance, the set of the first thousand positive integers may be specified extensionally as:

$$\{1, 2, 3, ..., 1000\},$$

where the ellipsis ("...") indicates that the list continues in the obvious way. Ellipses may also be used where sets have infinitely many members. Thus the set of positive even numbers can be written as $\{2, 4, 6, 8, ... \}$.

The notation with braces may also be used in an intensional specification of a set. In this usage, the braces have the meaning "the set of all ...". So, E = {playing card suits} is the set whose four members are ♠, ♦, ♥, and ♣ . A more general form of this is set-builder notation, through which, for instance, the set F of the twenty smallest integers that are four less than perfect squares can be denoted:

$$F = \{n^2 - 4 : n \text{ is an integer; and } 0 \leq n \leq 19\}.$$

In this notation, the colon (":") means "such that", and the description can be interpreted as "F is the set of all numbers of the form $n^2 - 4$, such that n is a whole number in the range from 0 to 19 inclusive." Sometimes the vertical bar ("|") is used instead of the colon.

One often has the choice of specifying a set intensionally or extensionally. In the examples above, for instance, $A = C$ and $B = D$.

Element (Mathematics)

In mathematics, an element, or member, of a set is any one of the distinct objects that make up that set.

Sets

Writing $A = \{1, 2, 3, 4\}$ means that the elements of the set A are the numbers 1, 2, 3 and 4. Sets of elements of A, for example $\{1, 2\}$, are subsets of A.

Sets can themselves be elements. For example consider the set $B = \{1, 2, \{3, 4$. The elements of B are *not* 1, 2, 3 and 4. Rather, there are only three elements of B, namely the numbers 1 and 2, and the set $\{3, 4\}$.

The elements of a set can be anything. For example, $C = \{$ red, green, blue $\}$, is the set whose elements are the colours red, green and blue.

Notation and Terminology

The relation "is an element of", also called set membership, is denoted by ". Writing

$$x \in A$$

means that "x is an element of A". Equivalent expressions are "x is a member of A", "x belongs to A", "x is in A" and "x lies in A". The expressions "A includes x" and "A contains x" are also used to mean set membership, however some authors use them to mean instead "x is a subset of A". Logician George Boolos strongly urged that "contains" be used for membership only and "includes" for the subset relation only.

The LaTeX command for this symbol is "\in".

The Mathematica command for this symbol is "\[Element]".

The negation of set membership is denoted by $\in$.

Cardinality of Sets

The number of elements in a particular set is a property known as cardinality, informally this is the size of a set. In the above examples the cardinality of the set A is 4, while the cardinality of either of the sets B and C is 3. An infinite set is a set with an infinite number of elements, while a finite set is a set with a finite number of elements. The above examples are examples of finite sets. An example of an infinite set is the set of natural numbers, N = { 1, 2, 3, 4, ... }.

Examples

Using the Sets Defined Above, Namely A = {1, 2, 3, 4 }, B = {1, 2, {3, 4 and C = { red, green, blue }:

- $2 \in A$
- $\{3,4\} \in B$
- {3,4} is a member of B
- Yellow $\notin C$
- The cardinality of D = { 2, 4, 8, 10, 12 } is finite and equal to 5.
- The cardinality of P = { 2, 3, 5, 7, 11, 13, ...} (the prime numbers) is infinite (this was proven by Euclid).

Subset

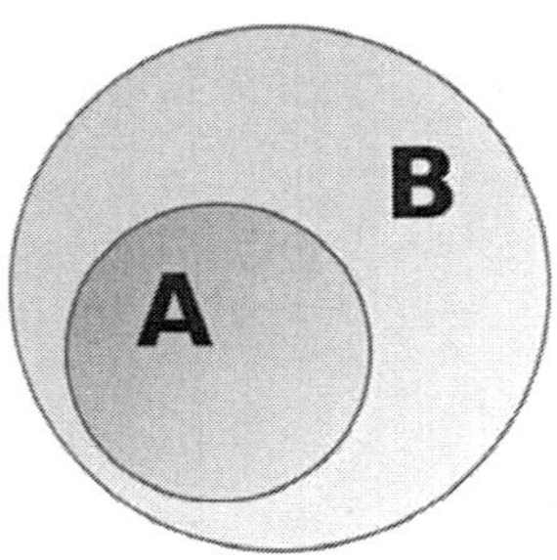

Figure: *Euler diagram showing* A *is a proper subset of* B *and conversely* B *is a proper superset of* A

In mathematics, especially in set theory, a set A is a subset of a set B, or equivalently B is a superset of A, if A is "contained" inside B, that is, all elements of A are also elements of B. A and B may coincide. The relationship of one set being a subset of another is called inclusion or sometimes containment.

Definitions

If A and B are sets and every element of A is also an element of B, then:

- A is a subset of (or is included in) B, denoted by $A \subseteq B$, or equivalently
- B is a superset of (or includes) A, denoted by $B \supseteq A$.

If A is a subset of B, but A is not equal to B (i.e. there exists at least one element of B not contained in A), then

- A is also a proper (or strict) subset of B; this is written as $A \subsetneq B$.

  or equivalently
- B is a proper superset of A; this is written as $B \supsetneq A$.

For any set S, the inclusion relation $\subseteq$ is a partial order on the set $\mathcal{P}(S)$ of all subsets of S (the power set of S).

The Symbols $\subset$ and f$\supset$

Some authors use the symbols $\subset$ and $\supset$ to indicate "subset" and "superset" respectively, instead of the symbols $\subseteq$ and $\supseteq$, but with the same meaning. So for example, for these authors, it is true of every set A that $A \subset A$.

Other authors prefer to use the symbols $\subset$ and $\supset$ to indicate *proper* subset and superset, respectively, in place of $\subsetneq$ and $\supsetneq$. This usage makes $\subseteq$ and $\subset$ analogous to the inequality symbols $\leq$ and $<$. For example, if $x \leq y$ then x may be equal to y, or maybe not, but if $x < y$, then x definitely does not equal y, and is strictly less than y. Similarly, using the "$\subset$ means proper subset" convention, if $A \subseteq B$, then A may or may not be equal to B, but if $A \subset B$, then A is definitely not equal to B.

Examples

- The set $\{1, 2\}$ is a proper subset of $\{1, 2, 3\}$.
- Any set is a subset of itself, but not a proper subset.

- The empty set { }, denoted by $\varnothing$, is also a subset of any given set X. It is also always a proper subset of any set except itself.
- The set {x: x is a prime number greater than 10} is a proper subset of {x: x is an odd number greater than 10}
- The set of natural numbers is a proper subset of the set of rational numbers, and the set of points in a line segment is a proper subset of the set of points in a line. These are examples in which both the part and the whole are infinite, and the part has the same cardinality (number of elements) as the whole; such cases can tax one's intuition.

Other Properties of Inclusion

Inclusion is the canonical partial order in the sense that every partially ordered set $(X, \preceq)$ is isomorphic to some collection of sets ordered by inclusion. The ordinal numbers are a simple example—if each ordinal n is identified with the set $[n]$ of all ordinals less than or equal to n, then $a \leq b$ if and only if $[a] \subseteq [b]$.

For the power set $\mathcal{P}(S)$ of a set S, the inclusion partial order is (up to an order isomorphism) the Cartesian product of $k = |S|$ (the cardinality of S) copies of the partial order on $\{0,1\}$ for which $0 < 1$. This can be illustrated by enumerating $S = \{s_1, s_2, ..., s_k\}$ and associating with each subset $T \subseteq S$ (which is to say with each element of 2^S) the k-tuple from $\{0,1\}^k$ of which the ith coordinate is 1 if and only if s_i is a member of T.

Power Set

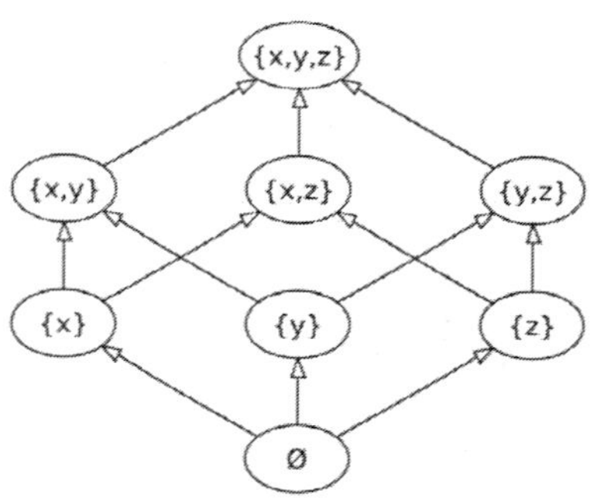

Figure: *The elements of the power set of the set {x, y, z} ordered in respect to inclusion.*

In mathematics, the power set (or powerset) of any set S, written $\mathcal{P}(S)$, $P(S)$, $\to(S)$, $\mathbb{P}(S)$ or 2^S, is the set of all subsets of S, including the empty set and S itself. In axiomatic set theory (as developed, for example, in the ZFC axioms), the existence of the power set of any set is postulated by the axiom of power set.

Any subset of $\mathcal{P}(S)$ is called a *family of sets* over S.

Example

If S is the set $\{x, y, z\}$, then the subsets of S are:

- (also denoted $\varnothing$, the empty set)
- $\{x\}$
- $\{y\}$
- $\{z\}$
- $\{x, y\}$
- $\{x, z\}$
- $\{y, z\}$
- $\{x, y, z\}$

and hence the power set of S is {, $\{x\}$, $\{y\}$, $\{z\}$, $\{x, y\}$, $\{x, z\}$, $\{y, z\}$, $\{x, y, z$.

Properties

If S is a finite set with $|S| = n$ elements, then the number of subsets of S is $|\mathcal{P}(S)| = 2^n$. This fact, which is the motivation for the notation 2^S, may be demonstrated by a simple inductive argument that illuminates the power set's combinatorial structure:

If n is zero, then S is the empty set, which has exactly $2^0 = 1$ subset (namely itself). Otherwise, n is some positive number, in which case one can select any element x of S and gather up the subsets of S itself into pairs (A^-, A^+) that are identical excepting that A^- does not include x and A^+ does. Thus, removing x from S would result in a new set whose own powerset—containing every A^- and nothing else—is exactly half the size of $\mathcal{P}(S)$. One can view S as having been built up from the empty set through successively restoring each of its

n elements, and during such a construction the power set undergoes n doublings in size. Hence $|\mathcal{P}(S)| = 2^n$.

Cantor's diagonal argument shows that the power set of a set (whether infinite or not) always has strictly higher cardinality than the set itself (informally the power set must be larger than the original set). In particular, Cantor's theorem shows that the power set of a countably infinite set is uncountably infinite. For example, the power set of the set of natural numbers can be put in a one-to-one correspondence with the set of real numbers.

The power set of a set S, together with the operations of union, intersection and complement can be viewed as the prototypical example of a Boolean algebra. In fact, one can show that any *finite* Boolean algebra is isomorphic to the Boolean algebra of the power set of a finite set. For *infinite* Boolean algebras this is no longer true, but every infinite Boolean algebra can be represented as a subalgebra of a power set Boolean algebra.

The power set of a set S forms an Abelian group when considered with the operation of symmetric difference (with the empty set as the identity element and each set being its own inverse) and a commutative monoid when considered with the operation of intersection. It can hence be shown (by proving the distributive laws) that the power set considered together with both of these operations forms a Boolean ring.

Representing Subsets as Functions

In set theory, X^Y is the set of all functions from Y to X. As "2" can be defined as $\{0,1\}$, 2^S (i.e., $\{0,1\}^S$) is the set of all functions from S to $\{0,1\}$. By identifying a function in 2^S with the corresponding preimage of 1, we see that there is a bijection between 2^S and $\mathcal{P}(S)$, where each function is the characteristic function of the subset in $\mathcal{P}(S)$ with which it is identified. Hence 2^S and $\mathcal{P}(S)$ could be considered identical set-theoretically. (Thus there are two distinct notational motivations for denoting the power set by 2^S: the fact that this function-representation of subsets makes it a special case of the X^Y notation and the property, mentioned above, that $|2^S| = 2^{|S|}$.)

This notion can be applied to the example above in which $S = \{x, y, z\}$ to see the isomorphism with the binary numbers from 0 to 2^{n}"1 with n being the number of elements in the set. In *S*, a *1* in the position corresponding to the location in the set indicates the presence of the element. So {*x*, *y*} = 110.

For the whole power set of *S* we get:

- { } = 000 (Binary) = 0 (Decimal)
- {*x*} = 100 = 4
- {*y*} = 010 = 2
- {*z*} = 001 = 1
- {*x*, *y*} = 110 = 6
- {*x*, *z*} = 101 = 5
- {*y*, *z*} = 011 = 3
- {*x*, *y*, *z*} = 111 = 7

Relation to Binomial Theorem

The power set is closely related to the binomial theorem. The number of sets with k elements in the power set of a set with n elements will be a combination $C(n, k)$, also called a binomial coefficient.

For example the power set of a set with three elements, has:

- $C(3,0) = 1$ set with 0 elements
- $C(3,1) = 3$ sets with 1 element
- $C(3,2) = 3$ sets with 2 elements
- $C(3,3) = 1$ set with 3 elements.

Algorithms

If *S*is a finite set, there is a recursive algorithm to calculate $\mathcal{P}(S)$.

Define the operation $\mathcal{F}(e, T) = \{X \cup \{e\} \mid X \in T\}$

In English, return the set with the element *e* added to each set *X*in *T*.

- If *S*=,then $\mathcal{P}(S) = \{\{\}\}$ is returned.
- Otherwise:

- Let e be any single element of S.
- Let $T = S \setminus \{e\}$, where '$S \setminus \{e\}$' denotes the relative complement of $\{e\}$ in S.
- And the result: $\mathcal{P}(S) = \mathcal{P}(T) \cup \mathcal{F}(e, \mathcal{P}(T))$ is returned.

In other words, the power set of the empty set is the set containing the empty set and the power set of any other set is all the subsets of the set containing some specific element and all the subsets of the set not containing that specific element.

There are other more efficient ways to calculate the power set. For example, use a list of the n elements of S to fix a mapping from the bit positions of n-bit numbers to those elements; then with a simple loop run through all the 2^n numbers representable with n bits, and for each contribute the subset of S corresponding to the bits that are set (to 1) in the number. When n exceeds the word-length of the computer (typically 64 in modern CPUs) the representation is naturally extended by using an array of words instead of a single word.

Subsets of Limited Cardinality

The set of subsets of S of cardinality less than κ is denoted by $\mathcal{P}_{\kappa}(S)$ or $\mathcal{P}_{<\kappa}(S)$. Similarly, the set of non-empty subsets of S might be denoted by $\mathcal{P}_{\geq 1}(S)$.

Power Object

A set can be regarded as an algebra having no nontrivial operations or defining equations. From this perspective the idea of the power set of X as the set of subsets of X generalizes naturally to the subalgebras of an algebraic structure or algebra.

Now the power set of a set, when ordered by inclusion, is always a complete atomic Boolean algebra, and every complete atomic Boolean algebra arises as the lattice of all subsets of some set. The generalization to arbitrary algebras is that the set of subalgebras of an algebra, again ordered by inclusion, is always an algebraic lattice, and every algebraic lattice arises as the lattice of subalgebras of some algebra. So in that regard subalgebras behave analogously to subsets.

However there are two important properties of subsets that do not carry over to subalgebras in general. First, although the subsets of a set form a set (as well as a lattice), in some classes it may not be possible to organize the subalgebras of an algebra as itself an algebra in that class, although they can always be organized as a lattice. Secondly, whereas the subsets of a set are in bijection with the functions from that set to the set $\{0,1\} = 2$, there is no guarantee that a class of algebras contains an algebra that can play the role of 2 in this way.

Certain classes of algebras enjoy both of these properties. The first property is more common, the case of having both is relatively rare. One class that does have both is that of multigraphs. Given two multigraphs G and H, a homomorphism $h: G \to H$ consists of two functions, one mapping vertices to vertices and the other mapping edges to edges. The set H^G of homomorphisms from G to H can then be organized as the graph whose vertices and edges are respectively the vertex and edge functions appearing in that set. Furthermore the subgraphs of a multigraph G are in bijection with the graph homomorphisms from G to the multigraph Ω definable as the complete directed graph on two vertices (hence four edges, namely two self-loops and two more edges forming a cycle) augmented with a fifth edge, namely a second self-loop at one of the vertices. We can therefore organize the subgraphs of G as the multigraph Ω^G, called the power object of G.

What is special about a multigraph as an algebra is that its operations are unary. A multigraph has two sorts of elements forming a set V of vertices and E of edges, and has two unary operations $s,t: E \to V$ giving the source (start) and target (end) vertices of each edge. An algebra all of whose operations are unary is called a presheaf. Every class of presheaves contains a presheaf Ω that plays the role for subalgebras that 2 plays for subsets. Such a class is a special case of the more general notion of elementary topos as a category that is closed (and moreover cartesian closed) and has an object Ω, called a subobject classifier. Although the term "power object" is sometimes used synonymously with exponential object Y^X, in topos theory Y is required to be Ω.

Cardinality

In mathematics, the cardinality of a set is a measure of the "number of elements of the set". For example, the set A =

{2, 4, 6} contains 3 elements, and therefore A has a cardinality of 3. There are two approaches to cardinality – one which compares sets directly using bijections and injections, and another which uses cardinal numbers.

The cardinality of a set A is usually denoted $| A |$, with a vertical bar on each side; this is the same notation as absolute value and the meaning depends on context. Alternatively, the cardinality of a set A may be denoted by n(A), A, card(A), or # A.

Comparing Sets

***Case 1:* $| A | = | B |$:** Two sets A and B have the same cardinality if there exists a bijection, that is, an injective and surjective function, from A to B.

For example, the set $E = \{0, 2, 4, 6, ...\}$ of non-negative even numbers has the same cardinality as the set N = {0, 1, 2, 3, ...} of natural numbers, since the function $f(n) = 2n$ is a bijection from N to E.

Case 2: | A | ≥ | B |

A has cardinality greater than or equal to the cardinality of B if there exists an injective function from B into A.

Case 3: | A | > | B |

A has cardinality strictly greater than the cardinality of B if there is an injective function, but no bijective function, from B to A.

For example, the set R of all real numbers has cardinality strictly greater than the cardinality of the set N of all natural numbers, because the inclusion map $i : \mathrm{N} \rightarrow \mathrm{R}$ is injective, but it can be shown that there does not exist a bijective function from N to R.

Cardinal Numbers

Above, “cardinality” was defined functionally. That is, the “cardinality” of a set was not defined as a specific object itself. However, such an object can be defined as follows.

The relation of having the same cardinality is called equinumerosity, and this is an equivalence relation on the class of all sets. The equivalence class of a set A under this relation then consists of all those sets which have the same cardinality as A. There are two ways to define the "cardinality of a set":

1. The cardinality of a set A is defined as its equivalence class under equinumerosity.
2. A representative set is designated for each equivalence class. The most common choice is the initial ordinal in that class. This is usually taken as the definition of cardinal number in axiomatic set theory.

The cardinalities of the infinite sets are denoted

$$\aleph_0 < \aleph_1 < \aleph_2 < \ldots.$$

For each ordinal α, $\aleph_{\alpha+1}$ is the least cardinal number greater than $\aleph_\alpha$.

The cardinality of the natural numbers is denoted aleph-null ($\aleph_0$), while the cardinality of the real numbers is denoted by c, and is also referred to as the cardinality of the continuum. Cantor showed, using the diagonal argument, that $c > \aleph_0$. We can show that $c = 2^{\aleph_0}$; this also being the cardinality of the set of all subsets of the natural numbers. The continuum hypothesis says that $\aleph_1 = 2^{\aleph_0}$, i.e. $2^{\aleph_0}$ is the smallest cardinal number bigger than $\aleph_0$, i.e. there is no set whose cardinality is strictly between that of the integers and that of the real numbers. The continuum hypothesis still remains unresolved in an "absolute" sense.

Finite, Countable and Uncountable Sets

If the axiom of choice holds, the law of trichotomy holds for cardinality. Thus we can make the following definitions:

- Any set X with cardinality less than that of the natural numbers, or $|X| < |N|$, is said to be a finite set.
- Any set X that has the same cardinality as the set of the natural numbers, or $|X| = |N| = \aleph_0$, is said to be a countably infinite set.
- Any set X with cardinality greater than that of the natural numbers, or $|X| > |N|$, for example $|R| = c > |N|$, is said to be uncountable.

Infinite Sets

Our intuition gained from finite sets breaks down when dealing with infinite sets. In the late nineteenth century Georg Cantor, Gottlob Frege, Richard Dedekind and others rejected the view of Galileo (which derived from Euclid) that the whole cannot be the same size as the part. One example of this is Hilbert's paradox of the Grand Hotel.

The reason for this is that the various characterizations of what it means for set A to be larger than set B, or to be the same size as set B, which are all equivalent for finite sets, are no longer equivalent for infinite sets. Different characterizations can yield different results. For example, in the popular characterization of size chosen by Cantor, sometimes an infinite set A is larger (in that sense) than an infinite set B; while other characterizations may yield that an infinite set A is always the same size as an infinite set B.

For finite sets, counting is just forming a bijection (i.e., a one-to-one correspondence) between the set being counted and an initial segment of the positive integers. Thus there is no notion equivalent to *counting* for infinite sets. While counting gives a unique result when applied to a finite set, an infinite set may be placed into a one-to-one correspondence with many different ordinal numbers depending on how one chooses to "count" (order) it.

Additionally, different characterizations of size, when extended to infinite sets, will break different "rules" which held for finite sets. Which rules are broken varies from characterization to characterization. For example, Cantor's characterization, while preserving the rule that sometimes one set is larger than another, breaks the rule that deleting an element makes the set smaller. Another characterization may preserve the rule that deleting an element makes the set smaller, but break another rule. Furthermore, some characterization may not "directly" break a rule, but it may not "directly" uphold it either, in the sense that whichever is the case depends upon a controversial axiom such as the axiom of choice or the continuum hypothesis. Thus there are three possibilities. Each characterization will break some rules, uphold some others, and may be indecisive about some others.

If one extends to multisets, further rules are broken (assuming Cantor's approach), which hold for finite multisets. If we have two multisets A and B, A not being larger than B and B not being larger than A does not necessarily imply A has the same size as B. This rule holds for multisets that are finite. Needless to say, the law of trichotomy is explicitly broken in this case, as opposed to the situation with sets, where it is equivalent to the axiom of choice.

Dedekind simply defined an infinite set as one having the same size (in Cantor's sense) as at least one of its proper parts; this notion of infinity is called Dedekind infinite. This definition only works in the presence of some form of the axiom of choice, however, so will not be considered to work by some mathematicians.

Cantor introduced the above-mentioned cardinal numbers, and showed that (in Cantor's sense) some infinite sets are greater than others. The smallest infinite cardinality is that of the natural numbers ($\aleph_0$).

Cardinality of the Continuum

One of Cantor's most important results was that the cardinality of the continuum ($\mathfrak{c}$) is greater than that of the natural numbers ($\aleph_0$); that is, there are more real numbers R than whole numbers N. Namely, Cantor showed that

$$\mathfrak{c} = 2^{\aleph_0} > \aleph_0$$

The continuum hypothesis states that there is no cardinal number between the cardinality of the reals and the cardinality of the natural numbers, that is,

$$\mathfrak{c} = \aleph_1 = \beth_1$$

However, this hypothesis can neither be proved nor disproved within the widely accepted ZFC axiomatic set theory, if ZFC is consistent.

Cardinal arithmetic can be used to show not only that the number of points in a real number line is equal to the number of points in any segment of that line, but that this is equal to the number of points on a plane and, indeed, in any finite-dimensional space. These results are highly counterintuitive, because they imply that there exist proper subsets and proper supersets of an infinite set S that have the same size as S, although S contains elements that do not belong to its subsets, and the supersets of S contain elements that are not included in it.

The first of these results is apparent by considering, for instance, the tangent function, which provides a one-to-one correspondence between the interval $(-\frac{1}{2}\pi, \frac{1}{2}\pi)$ and R.

The second result was first demonstrated by Cantor in 1878, but it became more apparent in 1890, when Giuseppe Peano introduced the space-filling curves, curved lines that twist and turn enough to fill the whole of any square, or cube, or hypercube, or finite-dimensional space. These curves are not a direct proof that a line has the same number of points as a finite-dimensional space, but they can be easily used to obtain such a proof.

Cantor also showed that sets with cardinality strictly greater than $\mathfrak{c}$ exist. They include, for instance:

- the set of all subsets of R, i.e., the power set of R, written $P(\mathrm{R})$ or 2^{R}
- the set R^{R} of all functions from R to R

Both have cardinality

$$2^{\mathfrak{c}} = \beth_2 > \mathfrak{c}$$

The cardinal equalities $\mathfrak{c}^2 = \mathfrak{c}$, $\mathfrak{c}^{\aleph_0} = \mathfrak{c}$, and $\mathfrak{c}^{\mathfrak{c}} = 2^{\mathfrak{c}}$ can be demonstrated using cardinal arithmetic:

$$\mathfrak{c}^2 = \left(2^{\aleph_0}\right)^2 = 2^{2\times\aleph_0} = 2^{\aleph_0} = \mathfrak{c},$$

$$\mathfrak{c}^{\aleph_0} = \left(2^{\aleph_0}\right)^{\aleph_0} = 2^{\aleph_0\times\aleph_0} = 2^{\aleph_0} = \mathfrak{c},$$

$$\mathfrak{c}^{\mathfrak{c}} = \left(2^{\aleph_0}\right)^{\mathfrak{c}} = 2^{\mathfrak{c}\times\aleph_0} = 2^{\mathfrak{c}}.$$

Examples and Properties

- If $X = \{a, b, c\}$ and $Y = \{$apples, oranges, peaches$\}$, then $X| = |Y|$ because $\{(a, \text{apples}), (b, \text{oranges}), (c, \text{peaches})\}$ is a bijection between the sets X and Y. The cardinality of each of X and Y is 3.

- If $|X| < |Y|$, then there exists Z such that $|X| = |Z|$ and $Z \subseteq Y$.
- If $|X| \leq |Y|$ and $|Y| \leq |X|$, then $|X| = |Y|$. This holds even for infinite cardinals, and is known as Cantor–Bernstein–Schroeder theorem.

Sets with cardinality of the continuum

Union and Intersection

If A and B are *disjoint* sets, then

$$|A \cup B| = |A| + |B|.$$

From this, one can show that in general the cardinalities of unions and intersections are related by

$$|C \cup D| + |C \cap D| = |C| + |D|.$$

Special Sets

There are some sets which hold great mathematical importance and are referred to with such regularity that they have acquired special names and notational conventions to identify them. One of these is the empty set, denoted or $\varnothing$. Another is the unit set {x} which contains exactly one element, namely x. Many of these sets are represented using blackboard bold or bold typeface. Special sets of numbers include:

- P or $\mathbb{P}$, denoting the set of all primes: P = {2, 3, 5, 7, 11, 13, 17, ...}.
- N or $\mathbb{N}$, denoting the set of all natural numbers: N = {1, 2, 3, . . .} (sometimes defined containing 0).
- Z or $\mathbb{Z}$, denoting the set of all integers (whether positive, negative or zero): Z = {..., –2, –1, 0, 1, 2, ...}.
- Q or $\mathbb{Q}$, denoting the set of all rational numbers (that is, the set of all proper and improper fractions): Q = $\{a/b : a, b \in Z, b \neq 0\}$. For example, $1/4 \in Q$ and $11/6 \in Q$. All integers are in this set since every integer a can be expressed as the fraction $a/1$ ($Z \subsetneq Q$).
- R or $\mathbb{R}$, denoting the set of all real numbers. This set includes all rational numbers, together with all irrational numbers (that is, numbers which cannot be rewritten as fractions, such as $\sqrt{2}$, as well as

transcendental numbers such as π, e and numbers that cannot be defined).

- C or $\mathbb{C}$, denoting the set of all complex numbers: C = $\{a + bi : a, b \in R\}$. For example, $1 + 2i \in C$.
- H or $\mathbb{H}$, denoting the set of all quaternions: H = $\{a + bi + cj + dk : a, b, c, d \in R\}$. For example, $1 + i + 2j \in k \in H$.

Positive and negative sets are denoted by a superscript - or +, for example: $\mathbb{Q}^+$ represents the set of positive rational numbers.

Each of the above sets of numbers has an infinite number of elements, and each can be considered to be a proper subset of the sets listed below it. The primes are used less frequently than the others outside of number theory and related fields.

Basic Operations

There are several fundamental operations for constructing new sets from given sets.

Union (Set Theory)

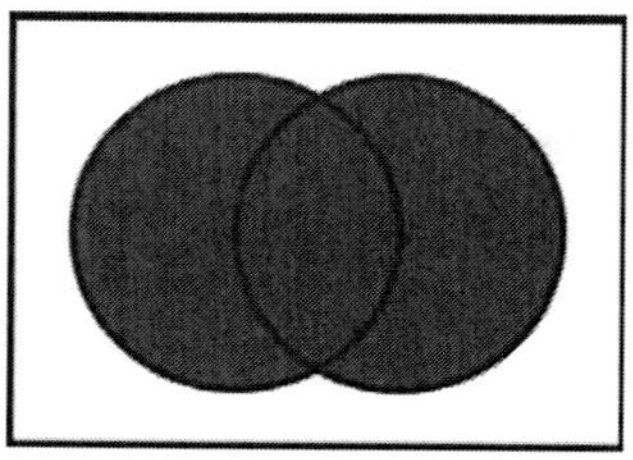

Figure: *Union of two sets:* $A \cup B$

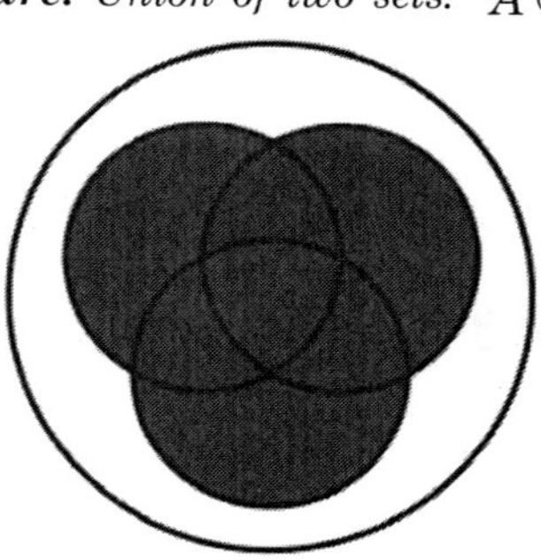

Figure: *Union of three sets:* $A \cup B \cup C$

In set theory, the union (denoted by $\cup$) of a collection of sets is the set of all distinct elements in the collection. It is one of the fundamental operations through which sets can be combined and related to each other.

Union of Two Sets

The union of two sets A and B is the collection of points which are in A or in B or in both A and B. In symbols,

$$A \cup B = \{x : x \in A \text{ or } x \in B\}.$$

For example, if $A = \{1, 3, 5, 7\}$ and $B = \{1, 2, 4, 6\}$ then $A \cup B = \{1, 2, 3, 4, 5, 6, 7\}$. A more elaborate example (involving two infinite sets) is:

$A = \{x$ is an even integer larger than $1\}$

$B = \{x$ is an odd integer larger than $1\}$

$$A \cup B = \{2,3,4,5,6,\ldots\}$$

If we are then to refer to a single element by the variable "x", then we can say that x is a member of the union if it is an element present in set A or in set B, or both.

Sets cannot have duplicate elements, so the union of the sets $\{1, 2, 3\}$ and $\{2, 3, 4\}$ is $\{1, 2, 3, 4\}$. Multiple occurrences of identical elements have no effect on the cardinality of a set or its contents. The number 9 is *not* contained in the union of the set of prime numbers $\{2, 3, 5, 7, 11, \ldots\}$ and the set of even numbers $\{2, 4, 6, 8, 10, \ldots\}$, because 9 is neither prime nor even.

Algebraic Properties

Binary union is an associative operation; that is,

$A \cup (B \cup C) = (A \cup B) \cup C$.

The operations can be performed in any order, and the parentheses may be omitted without ambiguity (i.e., either of the above can be expressed equivalently as $A \cup B \cup C$). Similarly, union is commutative, so the sets can be written in any order.

The empty set is an identity element for the operation of union. That is, $A \cup \varnothing = A$, for any set A.

These facts follow from analogous facts about logical disjunction.

Finite Unions

One can take the union of several sets simultaneously. For example, the union of three sets A, B, and C contains all elements of A, all elements of B, and all elements of C, and nothing else. Thus, x is an element of $A \cup B \cup C$ if and only if x is in at least one of A, B, and C.

In mathematics a finite union means any union carried out on a finite number of sets: it doesn't imply that the union set is a finite set.

Arbitrary Unions

The most general notion is the union of an arbitrary collection of sets, sometimes called an *infinitary union*. If M is a set whose elements are themselves sets, then x is an element of the union of M if and only if there is at least one element A of M such that x is an element of A. In symbols:

$$x \in \bigcup \mathrm{M} \Leftrightarrow \exists A \in \mathrm{M}, x \in A.$$

That this union of M is a set no matter how large a set M itself might be, is the content of the axiom of union in axiomatic set theory.

This idea subsumes the preceding sections, in that (for example) $A \cup B \cup C$ is the union of the collection $\{A,B,C\}$. Also, if M is the empty collection, then the union of M is the empty set. The analogy between finite unions and logical disjunction extends to one between arbitrary unions and existential quantification.

Notations

The notation for the general concept can vary considerably. For a finite union of sets $S_1, S_2, S_3, \ldots, S_n$ one often writes $S_1 \cup S_2 \cup S_3 \cup \ldots \cup S_n$. Various common notations for arbitrary unions include $\bigcup \mathrm{M}$, $\bigcup_{A \in \mathrm{M}} A$, and $\bigcup_{i \in I} A_i$, the last of which refers to the union of the collection $\{A_i : i \in I\}$ where I is an index set and A_i is a set for every $i \in I$. In the case that the index set I is the set of natural numbers, one uses a notation $\bigcup_{i=1}^{\infty} A_i$ analogous to that of the infinite series. When formatting is difficult, this can also be written "$A_1 \cup A_2 \cup A_3 \cup \cdots$". (This last example, a union of countably many sets, is very common in analysis.)

Whenever the symbol "$\cup$" is placed before other symbols instead of between them, it is of a larger size.

Union and Intersection

Intersection distributes over union, in the sense that

$$A \cap \bigcup_{i \in I} B_i = \bigcup_{i \in I} (A \cap B_i).$$

Within a given universal set, union can be written in terms of the operations of intersection and complement as

$$A \cup B = \left(A^C \cap B^C\right)^C$$

where the superscript ^{C} denotes the complement with respect to the universal set.

Arbitrary union and intersection also satisfy the law

$$\bigcup_{i \in I} \left(\bigcap_{j \in J} A_{i,j}\right) \subseteq \bigcap_{j \in J} \left(\bigcup_{i \in I} A_{i,j}\right).$$

Intersection (Set Theory)

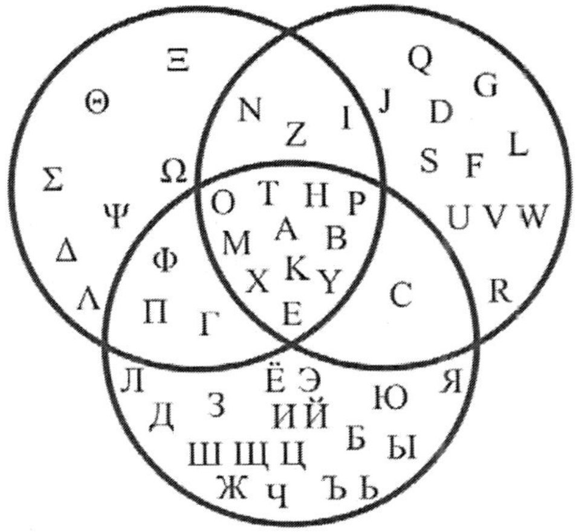

Figure: *Intersections of the Greek, Latin and Russian alphabet (upper case graphemes)*

In mathematics, the intersection (denoted as $\cap$) of two sets A and B is the set that contains all elements of A that also belong to B (or equivalently, all elements of B that also belong to A), but no other elements.

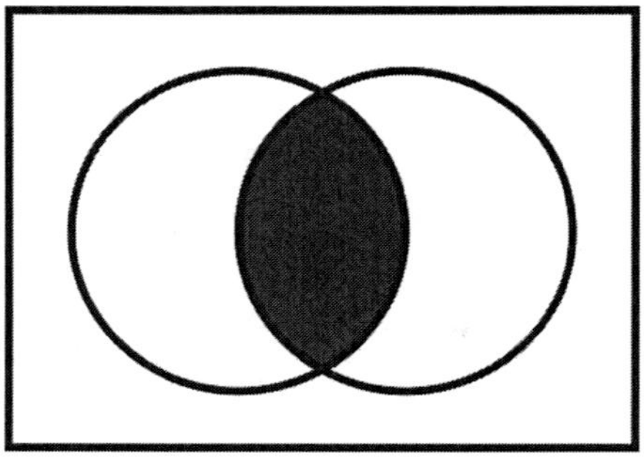

Figure: *Intersection of two sets:* $A \cap B$

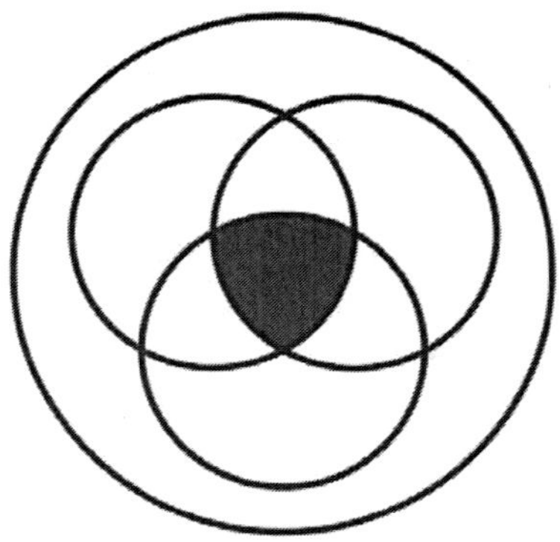

Figure: *Intersection of three sets:* $A \cap B \cap C$

For explanation of the symbols used in this article, refer to the table of mathematical symbols.

Basic Definition

The intersection of A and B is written "$A \cap B$". Formally:

$$A \cap B = \{x : x \in A \wedge x \in B\},$$

that is

$x \in A \cap B$ if and only if

- $x \in A$ and
- $x \in B$.

For example:

- The intersection of the sets {1, 2, 3} and {2, 3, 4} is {2, 3}.
- The number 9 is *not* in the intersection of the set of prime numbers {2, 3, 5, 7, 11, ...} and the set of odd numbers {1, 3, 5, 7, 9, 11, ...}.

More generally, one can take the intersection of several sets at once. The *intersection* of A, B, C, and D, for example, is $A \cap B \cap C \cap D = A \cap (B \cap (C \cap D))$. Intersection is an associative operation; thus,

$$A \cap (B \cap C) = (A \cap B) \cap C.$$

If the sets A and B are closed under complement then the intersection of A and B may be written as the complement of the union of their complements, derived easily from De Morgan's laws:

$$A \cap B = (A^c \cup B^c)^c$$

Intersecting and Disjoint Sets

We say that *A intersects (meets) B at an element x* if x belongs to A and B. We say that *A intersects (meets) B* if A intersects B at some element. A intersects B if their intersection is inhabited.

We say that *A and B are disjoint* if A does not intersect B. In plain language, they have no elements in common. A and B are disjoint if their intersection is empty, denoted $A \cap B = \varnothing$.

For example, the sets {1, 2} and {3, 4} are disjoint, the set of even numbers intersects the set of multiples of 3 at 0, 6, 12, 18 and other numbers.

Complement (Set Theory)

In set theory, a complement of a set A refers to things not in (that is, things outside of) A. The relative complement of A with respect to a set B, is the set of elements in B but not in A. When all sets under consideration are considered to be subsets of a given set U, the absolute complement of A is the set of all elements in U but not in A.

Relative Complement

If A and B are sets, then the relative complement of A in B, also termed the set-theoretic difference of B and A, is the set of elements in B, but not in A.

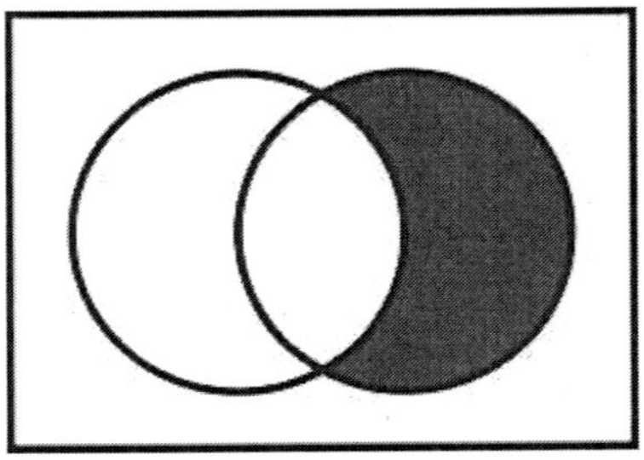

***Figure:** The relative complement of* A *(left circle) in* B *(right circle):*

$$A^c \cap B \;=\; B \setminus A$$

The relative complement of A in B is denoted $B \setminus A$ according to the ISO 31-11 standard (sometimes written $B - A$, but this notation is ambiguous, as in some contexts it can be interpreted as the set of all $b - a$, where b is taken from B and a from A).

Formally

$$B \setminus A = \{x \in B | x \notin A\}.$$

Examples:

- $\{1,2,3\} \setminus \{2,3,4\} \;=\; \{1\}$
- $\{2,3,4\} \setminus \{1,2,3\} \;=\; \{4\}$
- If $\mathbb{R}$ is the set of real numbers and $\mathbb{Q}$ is the set of rational numbers, then $\mathbb{R} \setminus \mathbb{Q} = \mathbb{J}$ is the set of irrational numbers.

The following lists some notable properties of relative complements in relation to the set-theoretic operations of union and intersection.

If A, B, and C are sets, then the following identities hold:

- $C \setminus (A \cap B) \;=\; (C \setminus A) \cup (C \setminus B)$
- $C \setminus (A \cup B) \;=\; (C \setminus A) \cap (C \setminus B)$
- $C \setminus (B \setminus A) \;=\; (A \cap C) \cup (C \setminus B)$
- $(B \setminus A) \cap C \;=\; (B \cap C) \setminus A \;=\; B \cap (C \setminus A)$
- $(B \setminus A) \cup C \;=\; (B \cup C) \setminus (A \setminus C)$
- $A \setminus A \;=\; \emptyset$
- $\emptyset \setminus A \;=\; \emptyset$
- $A \setminus \emptyset \;=\; A$

Absolute Complement

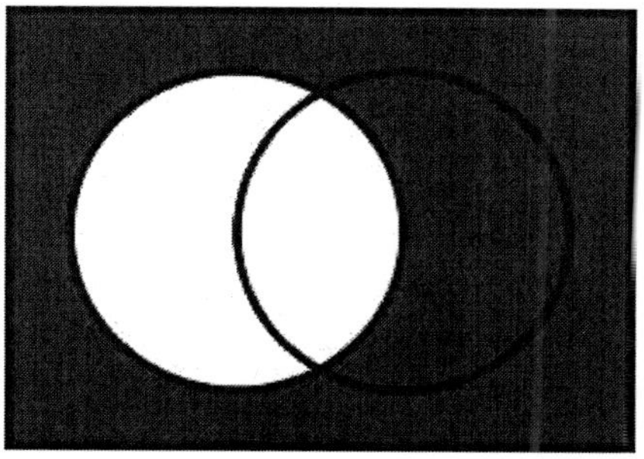

***Figure:** The complement of A in U:* $A^c = U \setminus A$

If a universe U is defined, then the relative complement of A in U is called the absolute complement (or simply complement) of A, and is denoted by A^c or sometimes $A2$, also the same set often is denoted by $\complement_U A$ or $\complement A$ if U is fixed, that is:

$A^c = U \setminus A.$

For example, if the universe is the set of integers, then the complement of the set of odd numbers is the set of even numbers.

The following lists some important properties of absolute complements in relation to the set-theoretic operations of union and intersection.

If A and B are subsets of a universe U, then the following identities hold:

De Morgan's laws:

- $(A \cup B)^c = A^c \cap B^c.$
- $(A \cap B)^c = A^c \cup B^c.$

Complement laws:

- $A \cup A^c = U.$
- $A \cap A^c = \varnothing.$
- $\varnothing^c = U.$
- $U^c = \varnothing.$
- If $A \subset B$, then $B^c \subset A^c$.

(this follows from the equivalence of a conditional with its contrapositive)

Involution or double complement law:

- $\left(A^c\right)^c = A.$

Relationships between relative and absolute complements:

- $A \setminus B = A \cap B^c$
- $(A \setminus B)^c = A^c \cup B$

The first two complement laws above shows that if A is a non-empty, proper subset of U, then $\{A, A^c\}$ is a partition of U.

Notation

In the LaTeX typesetting language, the command \setminus is usually used for rendering a set difference symbol, which is similar to a backslash symbol. When rendered the \setminus command looks identical to \backslash except that it has a little more space in front and behind the slash, akin to the LaTeX sequence \mathbin{\backslash}. A variant \smallsetminus is available in the amssymb package.

Complements in Various Programming Languages

Some programming languages allow for manipulation of sets as data structures, using these operators or functions to construct the difference of sets a and b:

.NET Framework

```
a.Except(b);
```

C++

```
set_difference(a.begin(), a.end(),
b.begin(), b.end(), result.begin());
```

Clojure

```
(clojure.set/difference a b)
```

Common Lisp

```
set-difference, nset-difference
```

Haskell

```
a \\ b
```

Java

```
diff = a.clone();
diff.removeAll(b);
```

Mathematica

```
Complement
```

MATLAB

```
setdiff
```

OCaml

```
Set.S.diff
```

Pascal

```
SetDifference := a - b;
```

Perl

```
#for perl version >= 5.10
@a = grep {not $_ ~~ @b} @a;
```

PHP

```
array_diff($a, $b);
```

Prolog

```
a(X),\+ b(X).
```

Python

```
diff = a.difference(b)
diff = a - b
```

R

```
setdiff
```

Ruby

```
diff = a - b
```

Scala

```
diff = a — b
```

SQL

```
SELECT * FROM A
MINUS
SELECT * FROM B
```

Unix shell

```
comm -23 a b
grep -vf b a # less efficient, but works
with small unsorted sets
```

Cartesian Product

In mathematics, a Cartesian product is a mathematical operation which returns a set (or product set) from multiple sets. That is, for sets A and B, the Cartesian product A × B is the set of all ordered pairs (a,b) where a ∈ A and b ∈ B.

The simplest case of a Cartesian product is the Cartesian square, which returns a set from two sets. A table can be created by taking the Cartesian product of a set of rows and a set of columns. If the Cartesian product rows × columns is taken, the cells of the table contain ordered pairs of the form (row value, column value). If the Cartesian product is columns × rows is taken, the cells of the table contain the ordered pairs of the form (column value, row value).

A Cartesian product of n sets can be represented by an array of n dimensions, where each element is an n-tuple. An ordered pair is a 2-tuple.

The Cartesian product is named after René Descartes, whose formulation of analytic geometry gave rise to the concept.

A two-dimensional Coordinate System

An example in analytic geometry is the Cartesian plane. The Cartesian plane is the result of the Cartesian product of two sets *X* and *Y*, which refer to points on the x-axis and points on the y-axis, respectively. This Cartesian product can be denoted as $X \times Y$. This produces the set of all possible ordered pairs whose first component is a member of *X* and whose second component is a member of *Y* (e.g., the whole of the x–y plane). Alternatively, the Cartesian product can be denoted as $Y \times X$, in which case the first component of the order pair is a member of *Y* and the second component of the ordered pair is a member of *X*. The Cartesian product is therefore not commutative.

$$X \times Y = \{(x, y) \mid x \in X \wedge y \in Y\}.$$

$$Y \times X = \{(y, x) \mid y \in Y \wedge x \in X\}.$$

$$X \times Y \neq Y \times X$$

Most Common Implementation (Set Theory)

A formal definition of the Cartesian product from set-theoretical principles follows from a definition of ordered pair.

The most common definition of ordered pairs, the Kuratowski definition, is $(x,y)=\{\{x\},\{x,y\}\}$. Note that, under this definition, $X\times Y\subseteq \mathcal{P}(\mathcal{P}(X\cup Y))$, where $\mathcal{P}$ represents the power set. Therefore, the existence of the Cartesian product of any two sets in ZFC follows from the axioms of pairing, union, power set, and specification. Since functions are usually defined as a special case of relations, and relations are usually defined as subsets of the Cartesian product, the definition of the two-set Cartesian product is necessarily prior to most other definitions.

Non-commutativity and Non-associativity

Let A, B, C, and D be sets.

The Cartesian product $A \times B$ is not commutative,

$$A\times B\neq B\times A,$$

because the ordered pairs are reversed except if at least one condition is satisfied:

- A is equal to B, or
- A or B is an empty set.

For example:

$A = \{1,2\}$; $B = \{3,4\}$

$A \times B = \{1,2\} \times \{3,4\} = \{(1,3), (1,4), (2,3), (2,4)\}$

$B \times A = \{3,4\} \times \{1,2\} = \{(3,1), (3,2), (4,1), (4,2)\}$

$A = B = \{1,2\}$

$A\times B = B\times A = \{1,2\} \times \{1,2\} = \{(1,1), (1,2), (2,1), (2,2)\}$

$A = \{1,2\}$; $B = \varnothing$

$A \times B = \{1,2\} \times \varnothing = \varnothing$

$B \times A = \varnothing \times \{1,2\} = \varnothing$

Strictly speaking, the Cartesian product is not associative (unless the above condition occurs).

$$(A\times B)\times C\neq A\times(B\times C)$$

Intersections, Unions, and Subsets

The Cartesian product acts nicely with respect to intersections.

$$(A\cap B)\times(C\cap D)=(A\times C)\cap(B\times D)$$

Notice that in most cases the above statement is not true if we replace intersection with union.

$$(A \cup B) \times (C \cup D) \neq (A \times C) \cup (B \times D)$$

Here are some rules demonstrating distributivity with other operators:

$$A \times (B \cap C) = (A \times B) \cap (A \times C),$$
$$A \times (B \cup C) = (A \times B) \cup (A \times C),$$
$$A \times (B \setminus C) = (A \times B) \setminus (A \times C),$$
$$(A \times B)^c = (A^c \times B^c) \cup (A^c \times B) \cup (A \times B^c).$$

Other properties related with subsets are:

$$\text{if } A \subseteq B \text{ then } A \times C \subseteq B \times C,$$
$$\text{if both } A, B \neq \varnothing \text{ then } A \times B \subseteq C \times D \Leftrightarrow A \subseteq C \wedge B \subseteq D.$$

Cardinality

The cardinality of a set is similar to the number of elements of the set. For a simpler example, defining two sets: $A = \{a, b\}$ and $B = \{5, 6\}$. Both set A and set B consist of two elements each. Their Cartesian product, written as $A \times B$, results in a new set which has the following elements:

$A \times B = \{(a,5), (a,6), (b,5), (b,6)\}$.

Each element of A is combined with each element of B. Each pair makes up one element of the output set. The number of values in each pair is equal to the number of sets whose cartesian product is being taken; hence 2 in this case. The cardinality of the result set is equal to the product of the cardinalities of all the input sets. That is,

$|A \times B| = |A| \cdot |B|$

and similarly

$|A \times B \times C| = |A| \cdot |B| \cdot |C|$

and so on.

The cardinality of $A \times B$ is also infinity if A or B is an infinite set.

Cartesian Square and Cartesian Power

The Cartesian square (or binary Cartesian product) of a set X is the Cartesian product $X^2 = X \times X$. An example is the

2-dimensional plane $R^2 = R \times R$ where R is the set of real numbers - all points (x,y) where x and y are real numbers.

Higher Powers of a Set

The cartesian power of a set X can be defined as:

$$X^n = \underbrace{X \times X \times \cdots \times X}_{n} = \{(x_1,\ldots,x_n) \mid x_i \in X \text{ for all } 1 \le i \le n\}.$$

An example of this is $R^3 = R \times R \times R$, with R again the set of real numbers, and more generally R^n.

Generalized Powers from Different sets

n-ary Product

The Cartesian product can be generalized to the n-ary Cartesian product over n sets $X_1, \ldots, X_n$:

$$X_1 \times \cdots \times X_n = \{(x_1,\ldots,x_n) : x_i \in X_i\}.$$

It is a set of n-tuples. If tuples are defined as nested ordered pairs, it can be identified to $(X_1 \times \ldots \times X_{n-1}) \times X_n$.

The n-ary cartesian power of a set X is isomorphic to the space of functions from an n-element set to X. As a special case, the 0-ary cartesian power of X may be taken to be a singleton set, corresponding to the empty function with codomain X.

Infinite Products

It is possible to define the Cartesian product of an arbitrary (possibly infinite) indexed family of sets. If I is any index set, and $\{X_i | i \in I\}$ is a collection of sets indexed by I, then the Cartesian product of the sets in X is defined to be

$$\prod_{i \in I} X_i = \left\{ f : I \to \bigcup_{i \in I} X_i \mid (\forall i)(f(i) \in X_i) \right\},$$

that is, the set of all functions defined on the index set such that the value of the function at a particular index i is an element of X_i. Even if each of the X_i is nonempty, the Cartesian product may be empty in general. The axiom of choice postulates that the product is nonempty.

For each j in I, the function

$$\pi_j : \prod_{i \in I} X_i \to X_j,$$

defined by $\pi_j(f) = f(j)$ is called the j-th projection map.

An important case is when the index set is $\mathbb{N}$, the natural numbers: this Cartesian product is the set of all infinite sequences with the i-th term in its corresponding set X_i. For example, each element of

$$\prod_{n=1}^{\infty}\mathbb{R} = \mathbb{R}\times\mathbb{R}\times\cdots$$

can be visualized as a vector with countably infinite real-number components. This set is frequently denoted $\mathbb{R}^{\omega}$, or $\mathbb{R}^{\mathbb{N}}$.

The special case Cartesian exponentiation occurs when all the factors X_i involved in the product are the same set X. In this case,

$$\prod_{i\in I} X_i = \prod_{i\in I} X$$

is the set of all functions from I to X, and is frequently denoted X^I. This case is important in the study of cardinal exponentiation.

The definition of finite Cartesian products can be seen as a special case of the definition for infinite products. In this interpretation, an n-tuple can be viewed as a function on {1, 2, ..., n} that takes its value at i to be the i-th element of the tuple (in some settings, this is taken as the very definition of an n-tuple).

Nothing in the definition of an infinite Cartesian product implies that the Cartesian product of nonempty sets must itself be nonempty. This assertion is equivalent to the axiom of choice.

Other Forms

Abbreviated Form: If several sets are being multiplied together, e.g. X_1, X_2, X_3, ..., then some authors choose to abbreviate the Cartesian product as simply $\times X_i$.

Cartesian Product of Functions

If f is a function from A to B and g is a function from X to Y, their cartesian product $f\times g$ is a function from $A\times X$ to $B\times Y$ with

$$(f\times g)(a,b) = (f(a), g(b)).$$

This can be extended to tuples and infinite collections of functions. Note that this is different from the standard cartesian product of functions considered as sets.

Definitions Outside of Set Theory

Category Theory: Although the Cartesian product is traditionally applied to sets, category theory provides a more general interpretation of the product of mathematical structures. This is distinct from, although related to, the notion of a Cartesian square in category theory, which is a generalization of the fibre product. Exponentiation is the right adjoint of the Cartesian product; thus any category with a Cartesian product (and a final object) is a Cartesian closed category.

Graph Theory

In graph theory the Cartesian product of two graphs G and H is the graph denoted by $G\times H$ whose vertex set is the (ordinary) Cartesian product $V(G)\times V(H)$ and such that two vertices (u,v) and $(u2, v2)$ are adjacent in $G\times H$ if and only if $u = v$ and u' is adjacent with v' in H, or $u' = v'$ and u is adjacent with v in G. The Cartesian product of graphs is not a product in the sense of category theory. Instead, the categorical product is known as the tensor product of graphs.

Inclusion–exclusion Principle

In combinatorics (combinatorial mathematics), the inclusion–exclusion principle is a counting technique which generalizes the familiar method of obtaining the number of elements in the union of two finite sets; symbolically expressed as

$$|A\cup B|=|A|+|B|-|A\cap B|,$$

where A and B are two finite sets and $|S|$ indicates the cardinality of a set S (which may be considered as the number of elements of the set, if the set is finite). The formula expresses the fact that if you just add the sizes of the two sets together, the sum will be too large since some elements will be counted twice and, to correct the result, you remove the double counting by subtracting the size of the intersection. The principle is more clearly seen in the case of three sets, which for the sets A, B and C is symbolically represented by

$$|A\cup B\cup C|=|A|+|B|+|C|-|A\cap B|-|A\cap C\ -|B\cap C|+|A\cap B\cap C|.$$

This formula can be verified by counting how many times each region in the Venn diagram figure is included in the right hand side of the formula. In this case, when removing the contributions of overcounted elements, the number of elements in the mutual intersection of the three sets has been subtracted too often, and so, must be added back in to get the correct total.

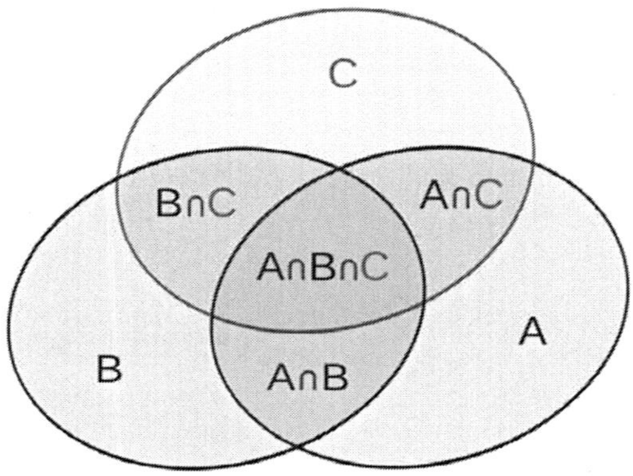

***Figure:** Inclusion–exclusion illustrated by a Venn diagram for three sets*

Generalizing the results of these examples gives the principle of inclusion-exclusion, to wit, to find the cardinality of the union of n sets, i) include the cardinalities of the sets, ii) exclude the cardinalities of the pairwise intersections, iii) include the cardinalities of the triplewise intersections, iv) exclude the cardinalities of the quadruple-wise intersections, v) include the cardinalities of the quintuple-wise intersections, vi) and continue, until you include (or exclude) the cardinality of the n-tuple-wise intersection.

The name comes from the idea that the principle is based on over-generous *inclusion*, followed by compensating *exclusion*. This concept is attributed to Abraham de Moivre (1718); but it first appears in a paper of Daniel da Silva (1854), and later in a paper by J. J. Sylvester (1883). Sometimes the principle is referred to as the formula of Da Silva, or Sylvester due to these publications. The principle is an example of the sieve method extensively used in number theory and is sometimes referred to as the *sieve formula*.

As finite probabilities are computed as counts relative to the cardinality of the probability space, the formulas for the principle of inclusion–exclusion, remain valid when the cardinalities of the sets are replaced by finite probabilites. More generally, both versions of the principle can be put under the common umbrella of measure theory.

In a very abstract setting, the principle of inclusion–exclusion amounts to no more than the calculation of the inverse of a certain matrix. From this point of view, there is nothing mathematically interesting about the principle. However, the wide applicability of the principle makes it an extremely valuable technique in combinatorics and related areas of mathematics. As Gian-Carlo Rota put it:

"One of the most useful principles of enumeration in discrete probability and combinatorial theory is the celebrated principle of inclusion-exclusion. When skillfully applied, this principle has yielded the solution to many a combinatorial problem."

Statement

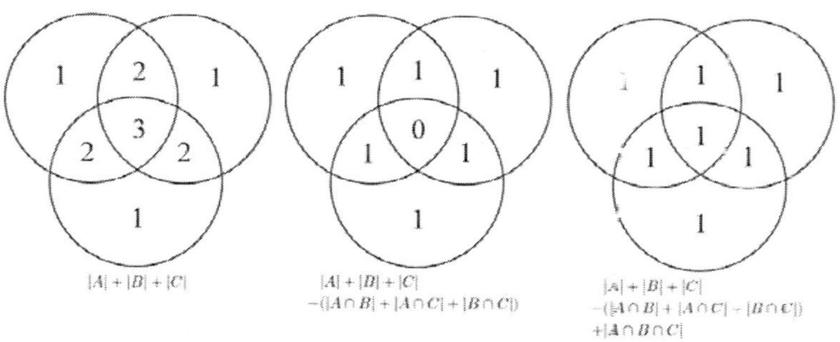

Figure: *Each term of the inclusion–exclusion formula gradually corrects the count until finally each portion of the Venn diagram is counted exactly once.*

In its general form, the principle of inclusion–exclusion states that for finite sets A_1, ..., A_n, one has the identity

$$\left|\bigcup_{i=1}^{n} A_i\right| = \sum_{i=1}^{n}|A_i| - \sum_{1\le i<j\le n}|A_i \cap A_j| + \sum_{1\le i<j<k\le n}|A_i \cap A_j \cap A_k| - \ldots + (-1)^{n-1}|A_1 \cap \cdots \cap A_n|.$$

This can be compactly written as

$$\left|\bigcup_{i=1}^{n} A_i\right| = \sum_{k=1}^{n}(-1)^{k+1}\left(\sum_{1\le i_1<\cdots<i_k\le n}\left|A_{i_1} \cap \cdots \cap A_{i_k}\right|\right).$$

In words, to count the number of elements in a finite union of finite sets, first sum the cardinalities of the individual sets, then subtract the number of elements which appear in more than one set, then add back the number of elements which appear in more than two sets, then subtract the number of elements which appear in more than three sets, and so on.

This process naturally ends since there can be no elements which appear in more than the number of sets in the union.

In applications it is common to see the principle expressed in its complementary form. That is, letting S be a finite universal set containing all of the A_i and letting $\overline{A_i}$ denote the complement of A_i in S, by De Morgan's laws we have

$$\left|\bigcap_{i=1}^{n}\overline{A_i}\right|=\left|S-\bigcup_{i=1}^{n}A_i\right|=|S|-\sum_{i=1}^{n}|A_i|+\sum_{1\le i<j\le n}|A_i\cap A_j|-\ldots+(-1)^n|A_1\cap\cdots\cap A_n|.$$

As another variant of the statement, let P_1, P_2, ..., P_n be a list of properties that elements of a set S may or may not have, then the principle of inclusion–exclusion provides a way to calculate the number of elements of S which have none of the properties. Just let A_i be the subset of elements of S which have the property P_i and use the principle in its complementary form. This variant is due to J.J. Sylvester.

Examples

As a simple example of the use of the principle of inclusion-exclusion, consider the question:

How many integers in {1,...,100} are not divisible by 2, 3 or 5?

Let S = {1,...,100} and P_1 the property that an integer is divisible by 2, P_2 the property that an integer is divisible by 3 and P_3 the property that an integer is divisible by 5. Letting A_i be the subset of S whose elements have property P_i we have by elementary counting: $|A_1| = 50$, $|A_2| = 33$, and $|A_3| = 20$. There are 16 of these integers divisible by 6, 10 divisible by 10 and 6 divisible by 15. Finally, there are just 3 integers divisible by 30, so the number of integers not divisible by any of 2, 3 or 5 is given by:

$$100 - (50 + 33 + 20) + (16 + 10 + 6) - 3 = 26.$$

A more complex example is the following.

Suppose there is a deck of n cards, each card is numbered from 1 to n. Suppose a card numbered m is in the correct position if it is the mth card in the deck. How many ways, W, can the cards be shuffled with at least 1 card being in the correct position?

Begin by defining set A_m, which is all of the orderings of cards with the mth card correct. Then the number of orders, W, with *at least* one card being in the correct position, m, is

$$W = \left| \bigcup_{m=1}^{n} A_m \right|.$$

Apply the principle of inclusion–exclusion,

$$\begin{aligned} W &= \sum_{m_1=1}^{n} \left|A_{m_1}\right| \\ &- \sum_{1 \le m_1 < m_2 \le n} \left|A_{m_1} \cap A_{m_2}\right| \\ &+ \sum_{1 \le m_1 < m_2 < m_3 \le n} \left|A_{m_1} \cap A_{m_2} \cap A_{m_3}\right| \\ &- \cdots \\ &+ (-1)^{p-1} \sum_{1 \le m_1 < \cdots < m_p \le n} \left|A_{m_1} \cap \cdots \cap A_{m_p}\right| \\ &\cdots. \end{aligned}$$

Each value $A_{m_1} \cap \cdots \cap A_{m_p}$ represents the set of shuffles having p values m_1, ..., m_p in the correct position. Note that the number of shuffles with p values correct only depends on p, not on the particular values of m. For example, the number of shuffles having the 1st, 3rd, and 17th cards in the correct position is the same as the number of shuffles having the 2nd, 5th, and 13th cards in the correct positions. It only matters that of the n cards, 3 were chosen to be in the correct position. Thus there are $\binom{n}{p}$ terms in each summation.

$$W = \binom{n}{1}|A_1| - \binom{n}{2}|A_1 \cap A_2| + \binom{n}{3}|A_1 \cap A_2 \cap A_3| - \cdots + (-1)^{p-1}\binom{n}{p}|A_1 \cap \cdots \cap A_p| \cdots.$$

$|A_1 \cap \cdots \cap A_p|$ is the number of orderings having p elements in the correct position, which is equal to the number of ways of ordering the remaining $n - p$ elements, or $(n - p)!$. Thus we finally get:

$$W = \binom{n}{1}(n-1)! - \binom{n}{2}(n-2)! + \binom{n}{3}(n-3)! - \cdots + (-1)^{p-1}\binom{n}{p}(n-p)! \cdots$$

$$W = \sum_{p=1}^{n}(-1)^{p-1}\binom{n}{p}(n-p)!.$$

Noting that $\binom{n}{p} = \dfrac{n!}{p!(n-p)!}$, this reduces to

$$W = \sum_{p=1}^{n}(-1)^{p-1}\frac{n!}{p!}.$$

A permutation where *no* card is in the correct position is called a derangement.

Taking $n!$ to be the total number of permutations, the probability Q that a random shuffle produces a derangement is given by

$$Q = 1 - \frac{W}{n!} = \sum_{p=0}^{n}\frac{(-1)^p}{p!},$$

a truncation to n+1 terms of the Taylor expansion of e^{-1}. Thus the probability of guessing an order for a shuffled deck of cards and being incorrect about every card is approximately $1/e$ or 37%.

A Special Case

The situation that appears in the derangement example above occurs often enough to merit special attention. Namely, when the size of the intersection sets appearing in the formulas for the principle of inclusion–exclusion depend only on the number of sets in the intersections and not on which sets appear. More formally, if the intersection

$$A_J := \bigcap_{j \in J} A_j$$

has the same cardinality, say $\alpha_k = |A_J|$, for every k-element subset J of $\{1, ..., n\}$, then

$$\left|\bigcup_{i=1}^{n} A_i\right| = \sum_{k=1}^{n}(-1)^{k-1}\binom{n}{k}\alpha_k.$$

Or, in the complementary form, where the universal set S has cardinality α_0,

$$\left|S\setminus\bigcup_{i=1}^{n} A_i\right| = \sum_{k=0}^{n}(-1)^{k}\binom{n}{k}\alpha_k.$$

A Generalization

Given a family (repeats allowed) of subsets $A_1, A_2, ..., A_n$ of a universal set S, the principle of inclusion-exclusion calculates the number of elements of S in none of these subsets. A generalization of this concept would calculate the number of elements of S which appear in exactly some fixed m of these sets.

Let $N = [n] = \{1,2,...,n\}$. If we define $A_{\varnothing} = S$, then the principle of inclusion-exclusion can be written as, using the notation of the previous section; the number of elements of S contained in none of the A_i is:

$$\sum_{J\subseteq[n]}(-1)^{|J|}\,|A_J|.$$

If I is a fixed subset of the index set N, then the number of elements which belong to A_i for all i in I and for no other values is:

$$\sum_{I\subseteq J}(-1)^{|J|-|I|}\,|A_J|.$$

Define the sets

$B_k = A_{I\cup\{k\}}$ for k in $N\setminus I$.

We seek the number of elements in none of the B_k which, by the principle of inclusion-exclusion (with $B_{\varnothing} = A_I$), is

$$\sum_{K\subseteq N\setminus I}(-1)^{|K|}\,|B_K|.$$

The correspondence $K \leftrightarrow J = I \cup K$ between subsets of $N \setminus I$ and subsets of N containing I is a bijection and if J and K correspond under this map then $B_K = A_J$, showing that the result is valid.

In Probability

In probability, for events A_1, ..., A_n in a probability space $(\Omega, \mathcal{F}, \mathbb{P})$, the inclusion–exclusion principle becomes for $n = 2$

$$\mathbb{P}(A_1 \cup A_2) = \mathbb{P}(A_1) + \mathbb{P}(A_2) - \mathbb{P}(A_1 \cap A_2),$$

for $n = 3$

$$\begin{aligned}\mathbb{P}(A_1 \cup A_2 \cup A_3) &= \mathbb{P}(A_1) + \mathbb{P}(A_2) + \mathbb{P}(A_3)\\ &\quad - \mathbb{P}(A_1 \cap A_2) - \mathbb{P}(A_1 \cap A_3) - \mathbb{P}(A_2 \cap A_3)\\ &\quad + \mathbb{P}(A_1 \cap A_2 \cap A_3)\end{aligned}$$

and in general

$$\begin{aligned}\mathbb{P}\Big(\bigcup_{i=1}^{n} A_i\Big) &= \sum_{i=1}^{n} \mathbb{P}(A_i) - \sum_{i<j} \mathbb{P}(A_i \cap A_j)\\ &\quad + \sum_{i<j<k} \mathbb{P}(A_i \cap A_j \cap A_k) - \cdots + (-1)^{n-1} \mathbb{P}\Big(\bigcap_{i=1}^{n} A_i\Big),\end{aligned}$$

which can be written in closed form as

$$\mathbb{P}\Big(\bigcup_{i=1}^{n} A_i\Big) = \sum_{k=1}^{n} (-1)^{k-1} \sum_{\substack{I \subset \{1,\ldots,n\} \\ |I|=k}} \mathbb{P}(A_I),$$

where the last sum runs over all subsets I of the indices 1, ..., n which contain exactly k elements, and

$$A_I := \bigcap_{i \in I} A_i$$

denotes the intersection of all those A_i with index in I.

According to the Bonferroni inequalities, the sum of the first terms in the formula is alternately an upper bound and a lower bound for the LHS. This can be used in cases where the full formula is too cumbersome.

For a general measure space (S, Σ, μ) and measurable subsets A_1, ..., A_n of finite measure, the above identities also hold when the probability measure $\mathbb{P}$ is replaced by the measure μ.

Special Case

If, in the probabilistic version of the inclusion–exclusion principle, the probability of the intersection A_I only depends on the cardinality of I, meaning that for every k in $\{1, ..., n\}$ there is an a_k such that

$$a_k = \mathbb{P}(A_I) \quad \text{for every} \quad I \subset \{1,\ldots,n\} \quad \text{with} \quad |I| = k,$$

then the above formula simplifies to

$$\mathbb{P}\Big(\bigcup_{i=1}^{n} A_i\Big) = \sum_{k=1}^{n} (-1)^{k-1} \binom{n}{k} a_k$$

due to the combinatorial interpretation of the binomial coefficient $\binom{n}{k}$.

An analogous simplification is possible in the case of a general measure space (S,Σ,μ) and measurable subsets A_1, ..., A_n of finite measure.

Applications

The inclusion–exclusion principle is widely used and only a few of its applications can be mentioned here.

Counting Derangements

A well-known application of the inclusion–exclusion principle is to the combinatorial problem of counting all derangements of a finite set. A *derangement* of a set A is a bijection from A into itself that has no fixed points. Via the inclusion–exclusion principle one can show that if the cardinality of A is n, then the number of derangements is $[n! / e]$ where $[x]$ denotes the nearest integer to x.

The first occurrence of the problem of counting the number of derangements is in an early book on games of chance: *Essai d'analyse sur les jeux de hazard* by P. R. de Montmort (1678 – 1719) and was known as either "Montmort's problem" or by the name he gave it, "*problème des rencontres*." The problem is also known as the *hatcheck problem.*

The number of derangements is also known as the subfactorial of n, written $!n$. It follows that if all bijections are assigned the same probability then the probability that a random bijection is a derangement quickly approaches $1/e$ as n grows.

Counting Intersections

The principle of inclusion–exclusion, combined with De Morgan's law, can be used to count the cardinality of the intersection of sets as well. Let $\overline{A}_k$ represent the complement of A_k with respect to some universal set A such that $A_k \subseteq A$ for each k. Then we have

$$\bigcap_{i=1}^{n} A_i = \overline{\bigcup_{i=1}^{n} \bar{A}_i}$$

thereby turning the problem of finding an intersection into the problem of finding a union.

Graph Colouring

The inclusion exclusion principle forms the basis of algorithms for a number of NP-hard graph partitioning problems, such as graph colouring.

A well known application of the principle is the construction of the chromatic polynomial of a graph.

Bipartite Graph Perfect Matchings

The number of perfect matchings of a bipartite graph can be calculated using the principle.

Number of Onto Functions

Given finite sets A and B, how many surjective functions (onto functions) are there from A to B? Without any loss of generality we may take $A = \{1,2,\ldots,k\}$ and $B = \{1,2,\ldots,n\}$, since only the cardinalities of the sets matter. By using S as the set of all functions from A to B, and defining, for each i in B, the property P_i as "the function misses the element i in B" (i is not in the image of the function), the principle of inclusion–exclusion gives the number of onto functions between A and B as:

$$\sum_{j=0}^{n} \binom{n}{j} (-1)^j (n-j)^k.$$

Permutations with Forbidden Positions

A permutation of the set $S = \{1,2,\ldots,n\}$ where each element of S is restricted to not being in certain positions (here the permutation is considered as an ordering of the elements of S) is called a *permutation with forbidden positions*. For example, with $S = \{1,2,3,4\}$, the permutations with the restriction that the element 1 can not be in positions 1 or 3, and the element 2 can not be in position 4 are: 2134, 2143, 3124, 4123, 2341,

2431, 3241, 3421, 4231 and 4321. By letting A_i be the set of positions that the element i is not allowed to be in, and the property P_i to be the property that a permutation puts element i into a position in A_i, the principle of inclusion-exclusion can be used to count the number of permutations which satisfy all the restrictions.

In the given example, there are 12 = 2(3!) permutations with property P_1, 6 = 3! permutations with property P_2 and no permutations have properties P_3 or P_4 as there are no restrictions for these two elements. The number of permutations satisfying the restrictions is thus:

4! - (12 + 6 + 0 + 0) + (4) = 24 - 18 + 4 = 10.

The final 4 in this computation is the number of permutations having both properties P_1 and P_2. There are no other non-zero contributions to the formula.

Stirling Numbers of the Second Kind

The Stirling numbers of the second kind, $S(n,k)$ count the number of partitions of a set of n elements into k non-empty subsets (indistinguishable *boxes*). An explicit formula for them can be obtained by applying the principle of inclusion-exclusion to a very closely related problem, namely, counting the number of partitions of an n-set into k non-empty but distinguishable boxes (ordered non-empty subsets). Using the universal set consisting of all partitions of the n-set into k (possibly empty) distinguishable boxes, A_1, A_2, ..., A_k, and the properties P_i meaning that the partition has box A_i empty, the principle of inclusion-exclusion gives an answer for the related result. Dividing by k! to remove the artificial ordering gives the Stirling number of the second kind:

$$S(n,k) = \frac{1}{k!}\sum_{t=0}^{k}(-1)^t \binom{k}{t}(k-t)^n.$$

Rook Polynomials

A rook polynomial is the generating function of the number of ways to place non-attacking rooks on a *board B* that looks like a subset of the squares of a checkerboard; that is, no two rooks may be in the same row or column. The board B is any subset of the squares of a rectangular board with n rows and

m columns; we think of it as the squares in which one is allowed to put a rook. The coefficient, $r_k(B)$ of x^k in the rook polynomial $R_B(x)$ is the number of ways k rooks, none of which attacks another, can be arranged in the squares of B. For any board B, there is a complementary board B' consisting of the squares of the rectangular board that are not in B. This complementary board also has a rook polynomial $R_{B'}(x)$ with coefficients $r_k(B')$.

It is sometimes convenient to be able to calculate the highest coefficient of a rook polynomial in terms of the coefficients of the rook polynomial of the complementary board. Without loss of generality we can assume that $n \leq m$, so this coefficient is $r_n(B)$. The number of ways to place n non-attacking rooks on the complete $n \times m$ "checkerboard" (without regard as to whether the rooks are placed in the squares of the board B) is given by the falling factorial:

$$(m)_n = m(m-1)(m-2)\cdots(m-n+1).$$

Letting P_i be the property that an assignment of n non-attacking rooks on the complete board has a rook in column i which is not in a square of the board B, then by the principle of inclusion-exclusion we have:

$$r_n(B) = \sum_{t=0}^{n}(-1)^t (m-t)_{n-t}\; r_t(B').$$

Euler's Phi Function

Euler's totient or phi function, $\varphi(n)$ is an arithmetic function that counts the number of positive integers less than or equal to n that are relatively prime to n. That is, if n is a positive integer, then $\varphi(n)$ is the number of integers k in the range $1 \leq k \leq n$ which have no common factor with n other than 1. The principle of inclusion-exclusion is used to obtain a formula for $\varphi(n)$. Let S be the set $\{1,2,\ldots,n\}$ and define the property P_i to be that a number in S is divisible by the prime number p_i, for $1 \leq i \leq r$, where the prime factorization of

$$n = p_1^{a_1} p_2^{a_2} \cdots p_r^{a_r}.$$

Then,

$$\phi(n) = n - \sum_{i=1}^{r}\frac{n}{p_i} + \sum_{1\leq i<j\leq r}\frac{n}{p_i p_j} - \cdots = n\prod_{i=1}^{r}\left(1-\frac{1}{p_i}\right).$$

Diluted Inclusion–exclusion Principle

In many cases where the principle could give an exact formula (in particular, counting prime numbers using the sieve of Eratosthenes), the formula arising doesn't offer useful content because the number of terms in it is excessive. If each term individually can be estimated accurately, the accumulation of errors may imply that the inclusion–exclusion formula isn't directly applicable. In number theory, this difficulty was addressed by Viggo Brun. After a slow start, his ideas were taken up by others, and a large variety of sieve methods developed. These for example may try to find upper bounds for the "sieved" sets, rather than an exact formula.

Let $A_1, ..., A_n$ be arbitrary sets and $p_1, ..., p_n$ real numbers in the closed unit interval [0,1]. Then, for every even number k in $\{0, ..., n\}$, the indicator functions satisfy the inequality:

$$1_{A_1\cup\cdots\cup A_n} \geq \sum_{j=1}^{k}(-1)^{j-1} \sum_{1\leq i_1<\cdots<i_j\leq n} p_{i_1}\cdots p_{i_j} 1_{A_{i_1}\cap\cdots\cap A_{i_j}}.$$

Proof

Let A denote the union $\cup_{i=1}^{n} A_i$ of the sets $A_1, ..., A_n$. To prove the inclusion–exclusion principle in general, we first have to verify the identity

$$1_A = \sum_{k=1}^{n}(-1)^{k-1} \sum_{\substack{I\subset\{1,\ldots,n\}\\|I|=k}} 1_{A_I} \qquad (*)$$

for indicator functions, where

$$A_I = \bigcap_{i\in I} A_i.$$

There are at least two ways to do this:

First possibility: It suffices to do this for every x in the union of $A_1, ..., A_n$. Suppose x belongs to exactly m sets with $1 \leq m \leq n$, for simplicity of notation say $A_1, ..., A_m$. Then the identity at x reduces to

$$1 = \sum_{k=1}^{m}(-1)^{k-1} \sum_{\substack{I\subset\{1,\ldots,m\}\\|I|=k}} 1.$$

The number of subsets of cardinality k of an m-element set is the combinatorical interpretation of the binomial coefficient $\binom{m}{k}$. Since $1 = \binom{m}{0}$, we have

$$\binom{m}{0} = \sum_{k=1}^{m} (-1)^{k-1} \binom{m}{k}.$$

Putting all terms to the left-hand side of the equation, we obtain the expansion for $(1-1)^m$ given by the binomial theorem, hence we see that (*) is true for x.

Second possibility: The following function is identically zero

$$(1_A - 1_{A_1})(1_A - 1_{A_2})\cdots(1_A - 1_{A_n}) = 0,$$

because: if x is not in A, then all factors are $0 - 0 = 0$; and otherwise, if x does belong to some A_m, then the corresponding mth factor is $1 - 1 = 0$. By expanding the product on the left-hand side, equation (*) follows.

Use of (*): To prove the inclusion–exclusion principle for the cardinality of sets, sum the equation (*) over all x in the union of A_1, ..., A_n. To derive the version used in probability, take the expectation in (*). In general, integrate the equation (*) with respect to μ. Always use linearity.

Alternative Proof

Pick an element contained in the union of all sets and let $A_1, A_2, \ldots, A_t$ be the individual sets containing it. (Note that $t > 0$.) Since the element is counted precisely once by the left-hand side of the equation, we need to show that it is counted precisely once by the right-hand side. By the binomial theorem,

$$(1-1)^t = \binom{t}{0} - \binom{t}{1} + \binom{t}{2} - \cdots + (-1)^t \binom{t}{t}.$$

Using the fact that $\binom{t}{0} = 1$ and rearranging terms, we have

$$\begin{aligned} 1 &= \binom{t}{1} - \binom{t}{2} + \cdots + (-1)^{t+1} \binom{t}{t} \\ &= |\{A_i \mid 1 \le i \le t\}| - |\{A_i \cap A_j \mid 1 \le i < j \le t\}| + \cdots + (-1)^{t+1} |\{A_1 \cap A_2 \cap \cdots \cap A_t\}|, \end{aligned}$$

and so the chosen element is indeed counted only once by the right-hand side of the proposed equation.

Algebra of Sets

The algebra of sets develops and describes the basic properties and laws of sets, the set-theoretic operations of union, intersection, and complementation and the relations of set equality and set inclusion. It also provides systematic procedures for evaluating expressions, and performing calculations, involving these operations and relations.

Introduction

The algebra of sets is the development of the fundamental properties of set operations and set relations. These properties provide insight into the fundamental nature of sets. They also have practical considerations.

Just like expressions and calculations in ordinary arithmetic, expressions and calculations involving sets can be quite complex. It is helpful to have systematic procedures available for manipulating and evaluating such expressions and performing such computations.

In the case of arithmetic, it is elementary algebra that develops the fundamental properties of arithmetic operations and relations.

For example, the operations of addition and multiplication obey familiar laws such as associativity, commutativity and distributivity; while the "less than or equal" relation satisfies such laws as reflexivity, antisymmetry and transitivity. These laws provide tools which facilitate computation as well as describe the fundamental nature of numbers, their operations and relations.

The algebra of sets is the set-theoretic analogue of the algebra of numbers. It is the algebra of the set-theoretic operations of union, intersection and complementation, and the relations of equality and inclusion. These are the topics covered in this article.

The Fundamental Laws of Set Algebra

The binary operations of set union and intersection satisfy many identities. Several of these identities or "laws" have well established names. Three pairs of laws, are stated, without proof, in the following proposition.

Proposition 1: For any sets *A*, *B*, and *C*, the following identities hold:

commutative laws:

- $A \cup B = B \cup A$
- $A \cap B = B \cap A$

associative laws:

- $(A \cup B) \cup C = A \cup (B \cup C)$
- $(A \cap B) \cap C = A \cap (B \cap C)$

distributive laws:

- $A \cup (B \cap C) = (A \cup B) \cap (A \cup C)$
- $A \cap (B \cup C) = (A \cap B) \cup (A \cap C)$

Notice that the analogy between unions and intersections of sets, and addition and multiplication of numbers, is quite striking. Like addition and multiplication, the operations of union and intersection are commutative and associative, and intersection *distributes* over unions. However, unlike addition and multiplication, union also *distributes* over intersection.

The next proposition, states two additional pairs of laws involving three specials sets: the empty set, the universal set and the complement of a set.

Proposition 2: For any subset *A* of universal set U, where Ø is the empty set, the following identities hold:

identity laws:

- $A \cup \varnothing = A$
- $A \cap U = A$

complement laws:

- $A \cup A^C = U$
- $A \cap A^C = \varnothing$

The identity laws (together with the commutative laws) say that, just like 0 and 1 for addition and multiplication, Ø and U are the identity elements for union and intersection, respectively.

Unlike addition and multiplication, union and intersection do not have inverse elements. However the complement laws give the fundamental properties of the somewhat inverse-like unary operation of set complementation.

The Preceding Five Pairs of Laws: the commutative, associative, distributive, identity and complement laws, can be said to encompass all of set algebra, in the sense that every valid proposition in the algebra of sets can be derived from them.

Note that if the complement laws are weakened to the rule $(A^C)^C = A$, then this is exactly the algebra of propositional linear logic.

The Principle of Duality

The above propositions display the following interesting pattern. Each of the identities stated above is one of a pair of identities such that each can be transformed into the other by interchanging $\cup$ and $\cap$, and also Ø and U.

These are examples of an extremely important and powerful property of set algebra, namely, the principle of duality for sets, which asserts that for any true statement about sets, the dual statement obtained by interchanging unions and intersections, interchanging U and Ø and reversing inclusions is also true. A statement is said to be self-dual if it is equal to its own dual.

Some Additional Laws for Unions and Intersections

The following proposition states six more important laws of set algebra, involving unions and intersections.

Proposition 3: For any subsets A and B of a universal set U, the following identities hold:

idempotent laws:

- $A \cup A = A$
- $A \cap A = A$

domination laws:

- $A \cup U = U$
- $A \cap \varnothing = \varnothing$

absorption laws:

- $A \cup (A \cap B) = A$
- $A \cap (A \cup B) = A$

As noted above each of the laws stated in proposition 3, can be derived from the five fundamental pairs of laws stated in proposition 1 and proposition 2. As an illustration, a proof is given below for the idempotent law for union.

Proof:

$A \cup A$	$= (A \cup A) \cap U$	by the identity law of intersection
	$= (A \cup A) \cap (A \cup A^C)$	by the complement law for union
	$= A \cup (A \cap A^C)$	by the distributive law of union over intersection
	$= A \cup \varnothing$	by the complement law for intersection
	$= A$	by the identity law for union

The following proof illustrates that the dual of the above proof is the proof of the dual of the idempotent law for union, namely the idempotent law for intersection.

Proof:

$A \cap A$	$= (A \cap A) \cup \varnothing$	by the identity law for union
	$= (A \cap A) \cup (A \cap A^C)$	by the complement law for intersection
	$= A \cap (A \cup A^C)$	by the distributive law of intersection over union
	$= A \cap U$	by the complement law for union
	$= A$	by the identity law for intersection

Intersection can be expressed in terms of union and set difference :

$$A \cap B = A \cup B - (A - B) - (B - A)$$

Some Additional Laws for Complements

The following proposition states five more important laws of set algebra, involving complements.

Proposition 4: Let A and B be subsets of a universe U, then:

De Morgan's laws:

- $(A \cup B)^C = A^C \cap B^C$
- $(A \cap B)^C = A^C \cup B^C$

double complement or Involution law:

- $(A^C)^C = A$

complement laws for the universal set and the empty set:

- $\varnothing^C = U$
- $U^C = \varnothing$

Notice that the double complement law is self-dual.

The next proposition, which is also self-dual, says that the complement of a set is the only set that satisfies the complement laws. In other words, complementation is characterized by the complement laws.

Proposition 5: Let A and B be subsets of a universe U, then:

uniqueness of complements:

- If $A \cup B = U$, and $A \cap B = \varnothing$, then $B = A^C$

The Algebra of Inclusion

The following proposition says that inclusion is a partial order.

PROPOSITION 6: If A, B and C are sets then the following hold:

reflexivity:

- $A \subseteq A$

antisymmetry:

- $A \subseteq B$ and $B \subseteq A$ if and only if $A = B$

transitivity:

- If $A \subseteq B$ and $B \subseteq C$, then $A \subseteq C$

The following proposition says that for any set S, the power set of S, ordered by inclusion, is a bounded lattice, and hence together with the distributive and complement laws above, show that it is a Boolean algebra.

Proposition 7: If A, B and C are subsets of a set S then the following hold:

existence of a least element and a greatest element:

- $\varnothing \subseteq A \subseteq S$

existence of joins:

- $A \subseteq A \cup B$
- If $A \subseteq C$ and $B \subseteq C$, then $A \cup B \subseteq C$

existence of meets:

- $A \cap B \subseteq A$
- If $C \subseteq A$ and $C \subseteq B$, then $C \subseteq A \cap B$

The following proposition says that the statement $A \subseteq B$ is equivalent to various other statements involving unions, intersections and complements.

Proposition 8: For any two sets A and B, the following are equivalent:

- $A \subseteq B$
- $A \cap B = A$
- $A \cup B = B$
- $A \setminus B = \varnothing$
- $B^C \subseteq A^C$

The above proposition shows that the relation of set inclusion can be characterized by either of the operations of set union or set intersection, which means that the notion of set inclusion is axiomatically superfluous.

The Algebra of Relative Complements

The following proposition lists several identities concerning relative complements or set-theoretic difference.

Proposition 9: For any universe U and subsets A, B, and C of U, the following identities hold:

- $C \setminus (A \cap B) = (C \setminus A) \cup (C \setminus B)$
- $C \setminus (A \cup B) = (C \setminus A) \cap (C \setminus B)$
- $C \setminus (B \setminus A) = (A \cap C) \cup (C \setminus B)$

- $(B \setminus A) \cap C = (B \cap C) \setminus A = B \cap (C \setminus A)$
- $(B \setminus A) \cup C = (B \cup C) \setminus (A \setminus C)$
- $A \setminus A = \varnothing$
- $\varnothing \setminus A = \varnothing$
- $A \setminus \varnothing = A$
- $B \setminus A = A^C \cap B$
- $(B \setminus A)^C = A \cup B^C$
- $U \setminus A = A^C$
- $A \setminus U = \varnothing$

Von Neumann Universe

In set theory and related branches of mathematics, the von Neumann universe, or von Neumann hierarchy of sets, denoted V, is the class of hereditary well-founded sets. This collection, which is formalized by Zermelo–Fraenkel set theory (ZFC), is often used to provide an interpretation or motivation of the axioms of ZFC.

The rank of a well-founded set is defined inductively as the smallest ordinal number greater than the ranks of all members of the set. In particular, the rank of the empty set is zero, and every ordinal has a rank equal to itself. The sets in V are divided into a transfinite hierarchy, called the cumulative hierarchy, based on their rank.

Definition

The cumulative hierarchy is a collection of sets V_α indexed by the class of ordinal numbers, in particular, V_α is the set of all sets having ranks less than α. Thus there is one set V_α for each ordinal number α; V_α may be defined by transfinite recursion as follows:

Let V_0 be the Empty Set,:

$$V_0 := \{\}.$$

- For any ordinal number β, let $V_{\beta+1}$ be the power set of V_β:

$$V_{\beta+1} := \mathcal{P}(V_\beta).$$

- For any limit ordinal λ, let V_λ be the union of all the V-stages so far:

$$V_\lambda := \bigcup_{\beta<\lambda} V_\beta.$$

A crucial fact about this definition is that there is a single formula $\varphi(\alpha,x)$ in the language of ZFC that defines "the set x is in V_α".

The class V is defined to be the union of all the V-stages:

$$V := \bigcup_{\alpha} V_\alpha.$$

An equivalent definition sets

$$V_\alpha := \bigcup_{\beta<\alpha} \mathcal{P}(V_\beta)$$

for each ordinal α, where $\mathcal{P}(X)$ is the powerset of X.

The rank of a set S is the smallest α such that $S \subseteq V_\alpha$.

V and Set Theory

If ω is the set of natural numbers, then V_ω is the set of hereditarily finite sets, which is a model of set theory without the axiom of infinity. $V_{\omega+\omega}$ is the universe of "ordinary mathematics", and is a model of Zermelo set theory.

If κ is an inaccessible cardinal, then V_κ is a model of Zermelo-Fraenkel set theory (ZFC) itself, and $V_{\kappa+1}$ is a model of Morse–Kelley set theory.

V is not "the set of all sets" for two reasons. First, it is not a set; although each individual stage V_α is a set, their union V is a proper class. Second, the sets in V are only the well-founded sets.

The axiom of foundation (or regularity) demands that every set is well founded and hence in V, and thus in ZFC every set is in V. But other axiom systems may omit the axiom of foundation or replace it by a strong negation (for example is Aczel's anti-foundation axiom). These non-well-founded set theories are not commonly employed, but are still possible to study.

Philosophical Perspectives

There are two approaches to understanding the relationship of the von Neumann universe *V* to ZFC (along with many variations of each approach, and shadings between them). Roughly, formalists will tend to view *V* as something that flows from the ZFC axioms (for example, ZFC proves that every set is in *V*). On the other hand, realists are more likely to see the von Neumann hierarchy as something directly accessible to the intuition, and the axioms of ZFC as propositions for whose truth in *V* we can give direct intuitive arguments in natural language. A possible middle position is that the mental picture of the von Neumann hierarchy provides the ZFC axioms with a motivation (so that they are not arbitrary), but does not necessarily describe objects with real existence.

Axiomatic Set Theory

Elementary set theory can be studied informally and intuitively, and so can be taught in primary schools using Venn diagrams. The intuitive approach tacitly assumes that a set may be formed from the class of all objects satisfying any particular defining condition. This assumption gives rise to paradoxes, the simplest and best known of which are Russell's paradox and the Burali-Forti paradox. Axiomatic set theory was originally devised to rid set theory of such paradoxes.

The most widely studied systems of axiomatic set theory imply that all sets form a cumulative hierarchy. Such systems come in two flavors, those whose ontology consists of:

- *Sets alone.* This includes the most common axiomatic set theory, Zermelo–Fraenkel set theory (ZFC), which includes the axiom of choice. Fragments of ZFC include:
  - Zermelo set theory, which replaces the axiom schema of replacement with that of separation;
  - General set theory, a small fragment of Zermelo set theory sufficient for the Peano axioms and finite sets;
  - Kripke-Platek set theory, which omits the axioms of infinity, powerset, and choice, and weakens the axiom schemata of separation and replacement.

- *Sets and proper classes*. These include Von Neumann–Bernays–Gödel set theory, which has the same strength as ZFC for theorems about sets alone, and Morse-Kelley set theory and Tarski–Grothendieck set theory, both of which are stronger than ZFC.

The above systems can be modified to allow urelements, objects that can be members of sets but that are not themselves sets and do not have any members.

The systems of New Foundations NFU (allowing urelements) and NF (lacking them) are not based on a cumulative hierarchy. NF and NFU include a "set of everything," relative to which every set has a complement. In these systems urelements matter, because NF, but not NFU, produces sets for which the axiom of choice does not hold.

Systems of constructive set theory, such as CST, CZF, and IZF, embed their set axioms in intuitionistic logic instead of first order logic. Yet other systems accept standard first order logic but feature a nonstandard membership relation. These include rough set theory and fuzzy set theory, in which the value of an atomic formula embodying the membership relation is not simply True or False. The Boolean-valued models of ZFC are a related subject.

An enrichment of ZFC called Internal Set Theory was proposed by Edward Nelson in 1977.

Applications

Many mathematical concepts can be defined precisely using only set theoretic concepts. For example, mathematical structures as diverse as graphs, manifolds, rings, and vector spaces can all be defined as sets satisfying various (axiomatic) properties. Equivalence and order relations are ubiquitous in mathematics, and the theory of mathematical relations can be described in set theory.

Set theory is also a promising foundational system for much of mathematics. Since the publication of the first volume of *Principia Mathematica*, it has been claimed that most or even all mathematical theorems can be derived using an aptly designed set of axioms for set theory, augmented with many definitions, using first or second order logic. For example, properties of the natural and real numbers can be derived within set theory, as each number system can be identified with a set of equivalence classes under a suitable equivalence relation whose field is some infinite set.

Set theory as a foundation for mathematical analysis, topology, abstract algebra, and discrete mathematics is likewise uncontroversial; mathematicians accept that (in principle) theorems in these areas can be derived from the relevant definitions and the axioms of set theory. Few full derivations of complex mathematical theorems from set theory have been formally verified, however, because such formal derivations are often much longer than the natural language proofs mathematicians commonly present. One verification project, Metamath, includes derivations of more than 10,000 theorems starting from the ZFC axioms and using first order logic.

Infinitary Combinatorics

In mathematics, infinitary combinatorics, or combinatorial set theory, is an extension of ideas in combinatorics to infinite sets. Some of the things studied include continuous graphs and trees, extensions of Ramsey's theorem, and Martin's axiom. Recent developments concern combinatorics of the continuum and combinatorics on successors of singular cardinals.

Ramsey Theory for Infinite Sets

Write κ, λ for ordinals, *m* for a cardinal number and *n* for a natural number. Erdõs & Rado (1956) introduced the notation

$$\kappa \to (\lambda)^n_m$$

as a shorthand way of saying that every partition of the set $[\kappa]^n$ of *n*-element subsets of *k* into *m* pieces has a homogeneous set of order type λ. A homogeneous set is in this case a subset of κ such that every *n*-element subset is in the same element of the partition. When *m* is 2 it is often omitted.

Assuming the Axiom of Choice, there are no ordinals κ with $\kappa\to(\omega)^\omega$, so *n* is usually taken to be finite. An extension where *n* is almost allowed to be infinite is the notation

$$\kappa \to (\lambda)^{<\omega}_m$$

which is a shorthand way of saying that every partition of the set of finite subsets of κ into *m* pieces has a subset of order type

λ such that for any finite n, all subsets of size n are in the same element of the partition. When m is 2 it is often omitted.

Another variation is the notation

$$\kappa \rightarrow (\lambda, \mu)^n$$

which is a shorthand way of saying that every colouring of the set $[\kappa]^n$ of n-element subsets of κ with 2 colours has a subset of order type λ such that all elements of $[\lambda]^n$ have the first colour, or a subset of order type μ such that all elements of $[\mu]^n$ have the second colour.

Some properties of this include: (in what follows k is a cardinal)

$\aleph_0 \rightarrow (\aleph_0)^n_k$ for all finite n and k (Ramsey's theorem).

$\beth_n^+ \rightarrow (\aleph_1)^{n+1}_{\aleph_0}$ (Erdõs–Rado theorem.)

$2^\kappa \nrightarrow (\kappa^+)^2$ (Sierpiñski theorem)

$2^\kappa \nrightarrow (3)^2_\kappa$

$\kappa \rightarrow (\kappa, \aleph_0)^2$ (Erdõs–Dushnik–Miller theorem).

In choiceless universes, partition properties with infinite exponents may hold, and some of them are obtained as consequences of the Axiom of determinacy (AD). For example, Donald A. Martin proved that AD implies

$$\aleph_1 \rightarrow (\aleph_1)^{\aleph_1}_2$$

Chapter 3

Descriptive Set Theory

In mathematical logic, descriptive set theory is the study of certain classes of "well-behaved" subsets of the real line and other Polish spaces. As well as being one of the primary areas of research in set theory, it has applications to other areas of mathematics such as functional analysis, ergodic theory, the study of operator algebras and group actions, and mathematical logic.

Polish Spaces

Descriptive set theory begins with the study of Polish spaces and their Borel sets.

A Polish space is a second countable topological space that is metrizable with a complete metric. Equivalently, it is a complete separable metric space whose metric has been "forgotten". Examples include the real line $\mathbb{R}$, the Baire space $\mathcal{N}$, the Cantor space $\mathcal{C}$, and the Hilbert cube $I^{\mathbb{N}}$.

Universality Properties

The class of Polish spaces has several universality properties, which show that there is no loss of generality in considering Polish spaces of certain restricted forms.

- Every Polish space is homeomorphic to a G_δ subspace of the Hilbert cube, and every G_δ subspace of the Hilbert cube is Polish.
- Every Polish space is obtained as a continuous image of Baire space; in fact every Polish space is the image of a continuous bijection defined on a closed subset of

Baire space. Similarly, every compact Polish space is a continuous image of Cantor space.

Because of these universality properties, and because the Baire space $\mathcal{N}$ has the convenient property that it is homeomorphic to $\mathcal{N}^{\omega}$, many results in descriptive set theory are proved in the context of Baire space alone.

Borel Sets

The class of Borel sets of a topological space X consists of all sets in the smallest σ-algebra containing the open sets of X. This means that the Borel sets of X are the smallest collection of sets such that:

- Every open subset of X is a Borel set.
- If A is a Borel set, so is $X \setminus A$. That is, the class of Borel sets are closed under complementation.
- If A_n is a Borel set for each natural number n, then the union $\bigcup A_n$ is a Borel set. That is, the Borel sets are closed under countable unions.

A fundamental result shows that any two uncountable Polish spaces X and Y are Borel isomorphic: there is a bijection from X to Y such that the preimage of any Borel set is Borel, and the image of any Borel set is Borel. This gives additional justification to the practice of restricting attention to Baire space and Cantor space, since these and any other Polish spaces are all isomorphic at the level of Borel sets.

Borel Hierarchy

Each Borel set of a Polish space is classified in the Borel hierarchy based on how many times the operations of countable union and complementation must be used to obtain the set, beginning from open sets. The classification is in terms of countable ordinal numbers. For each nonzero countable ordinal α there are classes Σ^0_α, Π^0_α, and Δ^0_α

- Every open set is declared to be Σ^0_1.
- A set is declared to be Π^0_α if and only if its complement is Σ^0_α.

- A set A is declared to be Σ^0_δ, $\delta > 1$, if there is a sequence)# A_i *# of sets, each of which is $\Pi^0_{\lambda(i)}$ for some $\lambda(i) < \delta$, such that $A = \bigcup A_i$.
- A set is Δ^0_α if and only if it is both Σ^0_α and Π^0_α.

A theorem shows that any set that is Σ^0_α or Π^0_α is $\Delta^0_{\alpha+1}$, and any Δ^0_β set is both Σ^0_α and Π^0_α for all $\alpha > \beta$. Thus the hierarchy has the following structure, where arrows indicate inclusion.

$$
\begin{array}{ccccccccccccc}
 & & \Sigma^0_1 & & & & \Sigma^0_2 & \cdots & & & \Sigma^0_\alpha & & \cdots \\
 & \nearrow & & \searrow & & \nearrow & & & & \nearrow & & \searrow & \\
\Delta^0_1 & & & & \Delta^0_2 & & & \cdots & \Delta^0_\alpha & & & & \Delta^0_{\alpha+1} \; \cdots \\
 & \searrow & & \nearrow & & \searrow & & & & \searrow & & \nearrow & \\
 & & \Pi^0_1 & & & & \Pi^0_2 & \cdots & & & \Pi^0_\alpha & & \cdots
\end{array}
$$

Regularity Properties of Borel Sets

Classical descriptive set theory includes the study of regularity properties of Borel sets. For example, all Borel sets of a Polish space have the property of Baire and the perfect set property. Modern descriptive set theory includes the study of the ways in which these results generalize, or fail to generalize, to other classes of subsets of Polish spaces.

Analytic and Coanalytic Sets

Just beyond the Borel sets in complexity are the analytic sets and coanalytic sets. A subset of a Polish space X is analytic if it is the continuous image of a Borel subset of some other Polish space. Although any continuous preimage of a Borel set is Borel, not all analytic sets are Borel sets. A set is coanalytic if its complement is analytic.

Projective Sets and Wadge Degrees

Many questions in descriptive set theory ultimately depend upon set-theoretic considerations and the properties of ordinal and cardinal numbers. This phenomenon is particularly apparent in the projective sets. These are defined via the projective hierarchy on a Polish space X:

- A set is declared to be Σ_1^1 if it is analytic.
- A set is Π_1^1 if it is coanalytic.
- A set A is Σ_{n+1}^1 if there is a Π_n^1 subset B of $X \times X$ such that A is the projection of B to the first coordinate.
- A set A is Π_{n+1}^1 if there is a Σ_n^1 subset B of $X \times X$ such that A is the projection of B to the first coordinate.
- A set Δ_n^1 is if it is both Π_n^1 and Σ_n^1.

As with the Borel hierarchy, for each n, any Δ_n^1 set is both Σ_{n+1}^1 and Π_{n+1}^1.

The properties of the projective sets are not completely determined by ZFC. Under the assumption $V = L$, not all projective sets have the perfect set property or the property of Baire. However, under the assumption of projective determinacy, all projective sets have both the perfect set property and the property of Baire. This is related to the fact that ZFC proves Borel determinacy, but not projective determinacy.

More generally, the entire collection of sets of elements of a Polish space X can be grouped into equivalence classes, known as Wadge degrees, that generalize the projective hierarchy. These degrees are ordered in the Wadge hierarchy. The axiom of determinacy implies that the Wadge hierarchy on any Polish space is well-founded and of length Θ, with structure extending the projective hierarchy.

Borel Equivalence Relations

A contemporary area of research in descriptive set theory studies Borel equivalence relations. A Borel equivalence relation on a Polish space X is a Borel subset of $X \times X$ that is an equivalence relation on X.

Effective Descriptive Set Theory

The area of effective descriptive set theory combines the methods of descriptive set theory with those of generalized recursion theory (especially hyperarithmetical theory). In particular, it focuses on lightface analogues of hierarchies of classical descriptive set theory. Thus the hyperarithmetic hierarchy is studied instead of the Borel hierarchy, and the analytical hierarchy instead of the projective hierarchy. This research is related to weaker version of set theory such as Kripke-Platek set theory and second-order arithmetic.

Fuzzy Set

Fuzzy sets are sets whose elements have degrees of membership. Fuzzy sets were introduced by Lotfi A. Zadeh and Dieter Klaua in 1965 as an extension of the classical notion of set. At the same time, Salii (1965) defined a more general kind of structures called *L*-relations, which were studied by him in an abstract algebraic context. Fuzzy relations, which are used now in different areas, such as linguistics (De Cock, et al, 2000), decision-making (Kuzmin, 1982) and clustering (Bezdek, 1978), are special cases of *L*-relations when *L* is the unit interval [0, 1].

In classical set theory, the membership of elements in a set is assessed in binary terms according to a bivalent condition — an element either belongs or does not belong to the set. By contrast, fuzzy set theory permits the gradual assessment of the membership of elements in a set; this is described with the aid of a membership function valued in the real unit interval [0, 1]. Fuzzy sets generalize classical sets, since the indicator functions of classical sets are special cases of the membership functions of fuzzy sets, if the latter only take values 0 or 1. In fuzzy set theory, classical bivalent sets are usually called *crisp* sets. The fuzzy set theory can be used in a wide range of domains in which information is incomplete or imprecise, such as bioinformatics.

It has been suggested by Thayer Watkins that Zadeh's ethnicity is an example of a fuzzy set because "His father was Turkish-Iranian (Azerbaijani) and his mother was Russian. His father was a journalist working in Baku, Azerbaijan in the Soviet Union...Lotfi was born in Baku in 1921 and lived there until his family moved to Tehran in 1931."

Fuzzy Logic

As an extension of the case of multi-valued logic, valuations ($\mu : V_o \to W$) of propositional variables (V_o) into a set of membership degrees (*W*) can be thought of as membership functions mapping predicates into fuzzy sets (or more formally, into an ordered set of fuzzy pairs, called a fuzzy relation). With these valuations, many-valued logic can be extended to allow for fuzzy premises from which graded conclusions may be drawn.

This extension is sometimes called "fuzzy logic in the narrow sense" as opposed to "fuzzy logic in the wider sense," which originated in the engineering fields of automated control and knowledge engineering, and which encompasses many topics involving fuzzy sets and "approximated reasoning."

Industrial applications of fuzzy sets in the context of "fuzzy logic in the wider sense" can be found at fuzzy logic.

Fuzzy Number

A fuzzy number is a convex, normalized fuzzy set $\widetilde{A} \subseteq \mathbb{R}$ whose membership function is at least segmentally continuous and has the functional value $\mu_A(x) = 1$ at precisely one element.

This can be likened to the funfair game "guess your weight," where someone guesses the contestant's weight, with closer guesses being more correct, and where the guesser "wins" if he or she guesses near enough to the contestant's weight, with the actual weight being completely correct (mapping to 1 by the membership function).

Fuzzy Interval

A fuzzy interval is an uncertain set $\widetilde{A} \subseteq \mathbb{R}$ with a mean interval whose elements possess the membership function value $\mu_A(x) = 1$. As in fuzzy numbers, the membership function must be convex, normalized, at least segmentally continuous.

Fuzzy Relation Equation

The fuzzy relation equation is an equation of the form A · R = B, where A and B are fuzzy sets, R is a fuzzy relation, and A · R stands for the composition of A with R.

Generalizations

There are many mathematical constructions similar to or more general than fuzzy sets. Since fuzzy sets were introduced in 1965, a lot of new mathematical constructions and theories treating imprecision, inexactness, ambiguity, and uncertainty have been developed. Some of these constructions and theories are extensions of fuzzy set theory, while others try to mathematically model imprecision and uncertainty in a different way (Burgin and Chunihin, 1997; Kerre, 2001; Deschrijver and Kerre, 2003).

Inner Model Theory

In set theory, inner model theory is the study of certain models of ZFC or some fragment or strengthening thereof. Ordinarily these models are transitive subsets or subclasses of the von Neumann universe *V*, or sometimes of a generic extension of *V*. Inner model theory studies the relationships of these models to determinacy, large cardinals, and descriptive set theory. Despite the name, it is considered more a branch of set theory than of model theory.

Examples

- The class of all sets is an inner model containing all other inner models.
- The first non-trivial example of an inner model was the constructible universe *L* developed by Kurt Gödel. Every model *M* of ZFC has an inner model L^M satisfying the axiom of constructibility, and this will be the smallest inner model of *M* containing all the ordinals of *M*. Regardless of the properties of the original model, L^M will satisfy the generalized continuum hypothesis and combinatorial axioms such as Ê%.
- The sets that are hereditarily ordinal definable form an inner model
- The sets that are hereditarily definable over a countable sequence of ordinals form an inner model, used in Solovay's theorem.

Consistency Results

One important use of inner models is the proof of consistency results. If it can be shown that every model of an axiom *A* has an inner model satisfying axiom *B*, then if *A* is consistent, *B* must also be consistent. This analysis is most useful when *A* is an axiom independent of ZFC, for example a large cardinal axiom; it is one of the tools used to rank axioms by consistency strength.

Large Cardinal

In the mathematical field of set theory, a large cardinal property is a certain kind of property of transfinite cardinal numbers. Cardinals with such properties are, as the name suggests, generally very "large" (for example, bigger than $\aleph_0$, bigger than the cardinality of the continuum, etc.). The proposition that such cardinals exist cannot be proved in the most common axiomatization of set theory, namely ZFC, and such propositions can be viewed as ways of measuring how "much", beyond ZFC, one needs to assume to be able to prove certain desired results. In other words, they can be seen, in Dana Scott's phrase, as quantifying the fact "that if you want more you have to assume more".

There is a rough convention that results provable from ZFC alone may be stated without hypotheses, but that if the proof requires other assumptions (such as the existence of large cardinals), these should be stated. Whether this is simply a linguistic convention, or something more, is a controversial point among distinct philosophical schools.

A large cardinal axiom is an axiom stating that there exists a cardinal (or perhaps many of them) with some specified large cardinal property.

There is no generally agreed precise definition of what a large cardinal property is, though essentially everyone agrees that those in the List of large cardinal properties are large cardinal properties.

Partial Definition

A necessary condition for a property of cardinal numbers to be a *large cardinal property* is that the existence of such a cardinal is not known to be inconsistent with ZFC and it has been proven that if ZFC is consistent, then ZFC + "no such cardinal exists" is consistent.

Hierarchy of Consistency Strength

A remarkable observation about large cardinal axioms is that they appear to occur in strict linear order by consistency strength. That is, no exception is known to the following: Given two large cardinal axioms A1 and A2, one of three (mutually exclusive) things happens:

1. ZFC proves "ZFC+A1 is consistent if and only if ZFC+A2 is consistent,"
2. ZFC+A1 proves that ZFC+A2 is consistent,
3. ZFC+A2 proves that ZFC+A1 is consistent.

In case 1 we say that A1 and A2 are equiconsistent. In case 2, we say that A1 is consistency-wise stronger than A2 (vice versa for case 3). If A2 is stronger than A1, then ZFC+A1 cannot prove A2 is consistent, even with the additional hypothesis that ZFC+A1 is itself consistent (provided of course that it really is). This follows from Gödel's second incompleteness theorem.

The observation that large cardinal axioms are linearly ordered by consistency strength is just that, an observation, not a theorem. (Without an accepted definition of large cardinal property, it is not subject to proof in the ordinary sense). Also, it is not known in every case which of the three cases holds. Saharon Shelah has asked, "[i]s there some theorem explaining this, or is our vision just more uniform than we realize?" Woodin, however, deduces this from the Ω-conjecture, the main unsolved problem of his Ω-logic.

It should also be noted that the order of consistency strength is not necessarily the same as the order of the size of the smallest witness to a large cardinal axiom. For example, the existence of a huge cardinal is much stronger, in terms of consistency strength, than the existence of a supercompact cardinal, but assuming both exist, the first huge is smaller than the first supercompact.

Motivations and Epistemic Status

Large cardinals are understood in the context of the von Neumann universe V, which is built up by transfinitely iterating the powerset operation, which collects together all subsets of a given set. Typically, models in which large cardinal axioms *fail* can be seen in some natural way as submodels of those in which the axioms hold. For example, if there is an inaccessible cardinal, then "cutting the universe off" at the height of the first such cardinal yields a universe in which there is no inaccessible cardinal. Or if there is a measurable cardinal, then iterating the *definable* powerset operation rather than the full one yields Gödel's constructible universe, L, which does not satisfy the statement "there is a measurable cardinal" (even though it contains the measurable cardinal as an ordinal).

Thus, from a certain point of view held by many set theorists (especially those inspired by the tradition of the Cabal), large cardinal axioms "say" that we are considering all the sets we're "supposed" to be considering, whereas their negations are "restrictive" and say that we're considering only some of those sets. Moreover the consequences of large cardinal axioms seem to fall into natural patterns. For these reasons, such set theorists tend to consider large cardinal axioms to have a preferred status among extensions of ZFC, one not shared by axioms of less clear motivation (such as Martin's axiom) or others that they consider intuitively unlikely (such as V = L). The hardcore realists in this group would state, more simply, that large cardinal axioms are *true*.

This point of view is by no means universal among set theorists. Some formalists would assert that standard set theory is by definition the study of the consequences of ZFC, and while they might not be opposed in principle to studying the consequences of other systems, they see no reason to single out large cardinals as preferred. There are also realists who deny that ontological maximalism is a proper motivation, and even believe that large cardinal axioms are false. And finally, there are some who deny that the negations of large cardinal axioms *are* restrictive, pointing out that (for example) there can be a transitive set model in L that believes there exists a measurable cardinal, even though L itself does not satisfy that proposition.

Determinacy

In set theory, a branch of mathematics, determinacy is the study of under what circumstances one or the other player of a game must have a winning strategy, and the consequences of the existence of such strategies.

Basic Notions

Games: The first sort of game we shall consider is the two-player game of perfect information of length ω, in which the players play natural numbers.

In this sort of game we consider two players, often named *I* and *II*, who take turns playing natural numbers, with *I* going first. They play "forever"; that is, their plays are indexed by the natural numbers. When they're finished, a predetermined condition decides which player won. This condition need not be specified by any definable *rule*; it may simply be an arbitrary (infinitely long) lookup table saying who has won given a particular sequence of plays.

More formally, consider a subset *A* of Baire space; recall that the latter consists of all ω-sequences of natural numbers. Then in the game G_A, *I* plays a natural number a_0, then *II* plays a_1, then *I* plays a_2, and so on. Then *I* wins the game if and only if

$$\langle a_0, a_1, a_2, \ldots \rangle \in A$$

and otherwise *II* wins. *A* is then called the *payoff set* of G_A.

It is assumed that each player can see all moves preceding each of his moves, and also knows the winning condition.

Strategies

Informally, a strategy for a player is a way of playing in which his plays are entirely determined by the foregoing plays. Again, such a "way" does not have to be capable of being captured by any explicable "rule", but may simply be a lookup table.

More formally, a strategy for player *I* (for a game in the sense of the preceding subsection) is a function that accepts as an argument any finite sequence of natural numbers, of even length, and returns a natural number. If σ is such a strategy and $<a_0,\ldots,a_{2n-1}>$ is a sequence of plays, then $\sigma(<a_0,\ldots,a_{2n-1}>)$ is the next play *I* will make, if he is following the strategy σ. Strategies for *II* are just the same, substituting "odd" for "even".

Note that we have said nothing, as yet, about whether a strategy is in any way *good*. A strategy might direct a player to make aggressively bad moves, and it would still be a strategy. In fact it is not necessary even to know the winning condition for a game, to know what strategies exist for the game.

Winning Strategies

A strategy is winning if the player following it must necessarily win, no matter what his opponent plays. For example if σ is a strategy for *I*, then σ is a winning strategy for *I* in the game G_A if, for any sequence of natural numbers to be played by *II*, say $<a_1,a_3,a_5,...>$, the sequence of plays produced by σ when *II* plays thus, namely

$$\langle\sigma(\langle\rangle),a_1,\sigma(\langle\sigma(\langle\rangle),a_1\rangle),a_3,\ldots\rangle$$

is an element of *A*.

Determined Games

A (class of) game(s) is determined if for all instance of the game there is a winning strategy for one of the players (not necessarily the same player for each instance). Note that there cannot be a winning strategy for *both* players for the same game, for if there were, the two strategies could be played against each other. The resulting outcome would then, by hypothesis, be a win for both players, which is impossible.

Determinacy from Elementary Considerations

All finite games of perfect information in which draws do not occur are determined. Familiar real-world games of perfect information, such as chess or tic-tac-toe, are always finished in a finite number of moves. If such a game is modified so that a particular player wins under any condition where the game would have been called a draw, then it is always determined. The condition that the game is always over (i.e. all possible extensions of the finite position result in a win for the same player) in a finite number of moves corresponds to the topological condition that the set *A* giving the winning condition for G_A is clopen in the topology of Baire space.

For example, modifying the rules of chess to make drawn games a win for Black makes chess a determined game. As it happens, chess has a finite number of positions and a draw-by-repetition rules, so with these modified rules, if play continues long enough without White having won, then Black can eventually force a win (due to the modification of draw = win for black).

It is an instructive exercise to figure out how to represent such games as games in the context of this article. The proof that such games are determined is rather simple: Player *I* simply plays *not to lose*; that is, he plays to make sure that player *II* does not have a winning strategy *after I*'s move. If player *I cannot* do this, then it means player *II* had a winning strategy from the beginning. On the other hand, if player *I can* play in this way, then he must win, because the game will be over after some finite number of moves, and he can't have lost at that point.

This proof does not actually require that the game *always* be over in a finite number of moves, only that it be over in a finite number of moves whenever *II* wins. That condition, topologically, is that the set *A* is closed. This fact—that all closed games are determined—is called the *Gale-Stewart theorem*. Note that by symmetry, all open games are determined as well. (A game is open if *I* can win only by winning in a finite number of moves.)

Determinacy from ZFC

Gale and Stewart proved the open and closed games are determined. Determinacy for second level of the Borel hierarchy games was shown by Wolfe in 1955. Over the following 20 years, additional research using ever-more-complicated arguments established that third and fourth levels of the Borel hierarchy are determined.

In 1975, Donald A. Martin proved that all Borel games are determined; that is, if *A* is a Borel subset of Baire space, then G_A is determined. This result, known as Borel determinacy, is the best possible determinacy result provable in ZFC, in the sense that the determinacy of the next higher Wadge class is not provable in ZFC.

In 1971, before Martin obtained his proof, Harvey Friedman showed that any proof of Borel determinacy must use the axiom of replacement in an essential way, in order to iterate the powerset axiom transfinitely often. Friedman's work gives a level-by-level result detailing how many iterations of the powerset axiom are necessary to guarantee determinacy at each level of the Borel hierarchy.

Determinacy and Large Cardinals

There is an intimate relationship between determinacy and large cardinals. In general, stronger large cardinal axioms prove the determinacy of larger pointclasses, higher in the Wadge hierarchy, and the determinacy of such pointclasses, in turn, proves the existence of inner models of slightly weaker large cardinal axioms than those used to prove the determinacy of the pointclass in the first place.

Measurable Cardinals

It follows from the existence of a measurable cardinal that every analytic game (also called a Σ_1^1 game) is determined, or equivalently that every coanalytic (or Π_1^1) game is determined.

Actually an apparently stronger result follows: If there is a measurable cardinal, then every game in the first ω^2 levels of the difference hierarchy over Π_1^1 is determined. This is only apparently stronger; ω^2- Π_1^1 determinacy turns out to be equivalent to Π_1^1 determinacy.

From the existence of more measurable cardinals, one can prove the determinacy of more levels of the difference hierarchy over Π_1^1.

Woodin Cardinals: If there is a Woodin cardinal with a measurable cardinal above it, then Π_2^1 determinacy holds. More generally, if there are n Woodin cardinals with a measurable cardinal above them all, then

Π_{n+1}^1 determinacy holds. From Π_{n+1}^1 determinacy, it follows that there is a transitive inner model containing n Woodin cardinals.

Projective Determinacy: If there are infinitely many Woodin cardinals, then projective determinacy holds; that is, every game whose winning condition is a projective set is determined. From projective determinacy it follows that, for every natural number n, there is a transitive inner model which satisfies that there are n Woodin cardinals.

Axiom of Determinacy: The axiom of determinacy, or AD, asserts that *every* two-player game of perfect information of length ω, in which the players play naturals, is determined.

AD is provably false from ZFC; using the axiom of choice one may prove the existence of a non-determined game. However, if there are infinitely many Woodin cardinals with a measurable above them all, then L(R) is a model of ZF that satisfies AD.

Consequences of Determinacy

Regularity Properties for Sets of Reals: If A is a subset of Baire space such that the Banach-Mazur game for A is determined, then either II has a winning strategy, in which case A is meager, or I has a winning strategy, in which case A is comeager on some open neighbourhood.

This does not quite imply that A has the property of Baire, but it comes close: A simple modification of the argument shows that if Γ is an adequate pointclass such that every game in Γ is determined, then every set of reals in Γ has the property of Baire.

In fact this result is not optimal; by considering the unfolded Banach-Mazur game we can show that determinacy of Γ (for Γ with sufficient closure properties) implies that every set of reals that is the *projection* of a set in Γ has the property of Baire. So for example the existence of a measurable cardinal implies Π^1_1 determinacy, which in turn implies that every Σ^1_2 set of reals has the property of Baire.

By considering other games, we can show that Π^1_n determinacy implies that every Σ^1_{n+1} set of reals has the property of Baire, is Lebesgue measurable (in fact universally measurable) and has the perfect set property.

Periodicity Theorems

- The first periodicity theorem implies that, for every natural number n, if Δ^1_{2n+1} determinacy holds, then Π^1_{2n+1} and Σ^1_{2n+2} have the prewellordering property (and that Σ^1_{2n+1} and Π^1_{2n+2} do *not* have the prewellordering property, but rather have the separation property).
- The second periodicity theorem implies that, for every natural number n, if Δ^1_{2n+1} determinacy holds, then Π^1_{2n+1} and Σ^1_{2n} have the scale property. In particular, if projective determinacy holds, then every projective relation has a projective uniformization.
- The third periodicity theorem gives a sufficient condition for a game to have a definable winning strategy.

Applications to Decidability of Certain Second-order Theories

In 1969, Michael O. Rabin proved that the second-order theory of n successors is decidable. A key component of the proof requires showing determinacy of parity games, which lie in the third level of the Borel hierarchy.

Wadge Determinacy

Wadge determinacy is the statement that for all pairs A,B of subsets of Baire space, the Wadge game G(A,B) is determined. Similarly for a pointclass Γ, Γ Wadge determinacy is the statement that for all sets A,B in Γ, the Wadge game G(A,B) is determined.

Wadge determinacy implies the semilinear ordering principle for the Wadge order. Another consequence of Wadge determinacy is the perfect set property.

In general, Γ Wadge determinacy is a consequence of the determinacy of Boolean combinations of sets in Γ. In the projective hierarchy, Π^1_1 Wadge determinacy is equivalent to Π^1_1 determinacy, as proved by Harrington. This result was extendend by Hjorth to prove that Π^1_2 Wadge determinacy (and in fact the semilinear ordering principle for Π^1_2) already implies Π^1_2 determinacy.

This subsection is still incomplete

More General Games

This section is still to be written

Games in Which the Objects Played are not Natural Numbers

This subsection is still to be written

Games Played on Trees

This subsection is still to be written

Long Games

This subsection is still to be written

Games of Imperfect Information

In any interesting game with imperfect information, a winning strategy will be a mixed strategy: that is, it will give some probability of differing responses to the same situation. If both players' optimal strategies are mixed strategies then the outcome of the game cannot be *certainly* determinant (as it can for pure strategies, since these are deterministic). But the probability distribution of outcomes to opposing mixed strategies can be calculated. A game that requires mixed strategies is defined as *determined* if a strategy exists that yields a minimum expected value (over possible counter-strategies) that exceeds a given value. Against this definition, all finite two player zero-sum games are clearly determined. However, the determinacy of *infinite* games of imperfect information (Blackwell games) is less clear.

In 1969 David Blackwell proved that some "infinite games with imperfect information" (now called "Blackwell games") are determined, and in 1998 Donald A. Martin proved that ordinary (perfect-information game) determinacy for a boldface pointclass implies Blackwell determinacy for the pointclass. This, combined with the Borel determinacy theorem of Martin, implies that all Blackwell games with Borel payoff functions are determined. Martin conjectured that ordinary determinacy and Blackwell determinacy for infinite games are equivalent in a strong sense (i.e. that Blackwell determinacy for a boldface pointclass in turn implies ordinary determinacy for that pointclass), but as of 2010, it has not been proven that Blackwell determinacy implies perfect-information-game determinacy.

Forcing (Mathematics)

In the mathematical discipline of set theory, forcing is a technique invented by Paul Cohen for proving consistency and independence results. It was first used, in 1963, to prove the independence of the axiom of choice and the continuum hypothesis from Zermelo–Fraenkel set theory. Forcing was considerably reworked and simplified in the 1960s, and has proven to be an extremely powerful technique both within set theory and in areas of mathematical logic such as recursion theory.

Descriptive set theory uses both the notion of forcing from recursion theory as well as set theoretic forcing. Forcing has also been used in model theory but it is common in model theory to define genericity directly without mention of forcing.

Intuitions

Forcing is equivalent to the method of Boolean-valued models, which some feel is conceptually more natural and intuitive, but usually much more difficult to apply.

Intuitively, forcing consists of expanding the set theoretical universe V to a larger universe V^*. In this bigger universe, for example, one might have lots of new subsets of $\omega = \{0,1,2,\ldots\}$ that were not there in the old universe, and thereby violate the continuum hypothesis. While impossible on the face of it, this is just another version of Cantor's paradox about infinity. In principle, one could consider

$$V^* = V \times \{0,1\},$$

identify $x \in V$ with $(x,0)$, and then introduce an expanded membership relation involving the "new" sets of the form $(x,1)$. Forcing is a more elaborate version of this idea, reducing the expansion to the existence of one new set, and allowing for fine control over the properties of the expanded universe.

Cohen's original technique, now called ramified forcing, is slightly different from the unramified forcing expounded here.

Forcing Posets

A forcing poset is an ordered triple

$$(\mathrm{P}, \leq, 1)$$

where "$\leq$" is a preorder on P that satisfies following splitting condition

For all $p \in \mathrm{P}$, there are $q, r \in \mathrm{P}$ such that $q, r \leq p$ with no $s \in \mathrm{P}$ such that $s \leq q, r$ and 1 is a largest element, that is,

$$p \leq 1 \text{ for all } p \in \mathrm{P}.$$

Members of P are called *conditions*. One reads

$$p \leq q$$

as

p is *stronger* than q.

Intuitively, the "smaller" condition provides "more" information, just as the smaller interval [3.1415926,3.1415927] provides more information about the number π than the interval [3.1,3.2] does.

(There are various conventions here. Some authors require "≤" to also be antisymmetric, so that the relation is a partial order. Some use the term partial order anyway, conflicting with standard terminology, while some use the term preorder. The largest element can be dispensed with. The reverse ordering is also used, most notably by Saharon Shelah and his co-authors.)

Associated with a forcing poset P are the P-*names*. P-names are sets of the form {(u,p):u is a P-name and p " P and (some criterion involving u and p)}.

This definition is circular; which in set theory means it is really a definition by transfinite recursion. In long form, one defines

- Name(0) =;
- Name(α + 1) = a well-defined subset of the power set of (Name(α) × P);
- Name(λ) = $\cup${Name(α) : $\alpha < \lambda$}, for λ a limit ordinal, and then defines the class of P-names to be

  $V^{(P)} = \cup\{\text{Name}(\alpha) : \alpha \text{ is an ordinal}\}$.

The P-names are, in fact, an expansion of the universe. Given x in V, one defines x¡ to be the P-name {(y¡,1) : $y \in x$}.

Again, this is really a definition by transfinite recursion.

Given any subset G of P, one next defines the *interpretation* or *valuation* map from names by

$$\text{val}(u, G) = \{\text{val}(v, G) : \exists\ p \in G\ ,\ (v, p) \in u\}.$$

(Again a definition by transfinite recursion.) Note that if 1 is in G, then val(x¡, G) = x.

One defines

$$G = \{(p¡, p) : p \in G\},$$

then

$$\text{val}(G,G) = G.$$

A good example of a forcing poset is

$$(\text{Bor(I)}, \subseteq, \text{I}),$$

where I = [0,1] and Bor(I) are the Borel subsets of I having non-zero Lebesgue measure. In this case, one can talk about the conditions as being probabilities, and a Bor(I)-name assigns membership in a probabilistic sense. Because of the ready intuition this example can provide, probabilistic language is sometimes used with other forcing posets.

Countable Transitive Models and Generic Filters

The key step in forcing is, given a ZFC universe V, to find appropriate G not in V. The resulting class of all interpretations of P-names will turn out to be a model of ZFC, properly extending the original V (since $G \notin V$).

Instead of working with V, one considers a countable transitive model M with $(P,\leq,1) \in M$. By model, we mean a model of set theory, either of all of ZFC, or a model of a large but finite subset of the ZFC axioms, or some variant thereof. Transitivity means that if $x \in y \in M$, then $x \in M$. The Mostowski collapsing theorem says this can be assumed if the membership relation is well-founded. The effect of transitivity is that membership and other elementary notions can be handled intuitively. Countability of the model relies on the Löwenheim–Skolem theorem.

Since M is a set, there are sets not in M – this follows from Russell's paradox. The appropriate set G to pick, and adjoin to M, is a generic filter on P. The *filter* condition means that $G \subseteq P$ and

- $1 \in G$;
- if $p \geq q \in G$, then $p \in G$;
- if $p,q \in G$, then $\exists r \in G$, $r \leq p$ and $r \leq q$;

For G to be *generic* means

- if $D \in M$ is a *dense* subset of P (that is, $p \in P$ implies $\exists q \in D$, $q \leq p$) then $G \cap D \neq 0$.

The existence of a generic filter G follows from the Rasiowa–Sikorski lemma. In fact, slightly more is true: given a condition $p \in P$, one can find a generic filter G such that $p \in G$. Due to the splitting condition, if G is filter, then $P \setminus G$ is dense. If G is in M then $P \setminus G$ is in M because M is model of set theory. By this reason, generic filter is never in M.

Consistency

The above can be summarized by saying the fundamental consistency result is that given a forcing poset P, we may assume that there exists a generic filter G, not in the universe V, such that $V[G]$ is again a set theoretic universe, modelling ZFC. Furthermore, all truths in $V[G]$ can be reduced to truths in V regarding the forcing relation.

Both styles, adjoining G to a countable transitive model M or to the whole universe V, are commonly used. Less commonly seen is the approach using the "internal" definition of forcing, and no mention of set or class models is made. This was Cohen's original method, and in one elaboration, it becomes the method of Boolean-valued analysis.

Cohen Forcing

The simplest nontrivial forcing poset is $(\mathrm{Fin}(\omega,2), \supseteq, 0)$, the finite partial functions from ω to $2=\{0,1\}$ under *reverse* inclusion. That is, a condition p is essentially two disjoint finite subsets $p^{-1}[1]$ and $p^{-1}[0]$ of ω, to be thought of as the "yes" and "no" parts of p, with no information provided on values outside the domain of p. q is stronger than p means that $q \supseteq p$, in other words, the "yes" and "no" parts of q are supersets of the "yes" and "no" parts of p, and in that sense, provide more information.

Let G be a generic filter for this poset. If p and q are both in G, then $p \cup q$ is a condition, because G is a filter. This means that $g=\cup G$ is a well-defined partial function from ω to 2, because any two conditions in G agree on their common domain.

g is in fact a total function. Given $n \in \omega$, let $D_n=\{ p : p(n)$ is defined $\}$, then D_n is dense. (Given any p, if n is not in p's domain, adjoin a value for n, the result is in D_n.) A condition $p \in G \cap D_n$ has n in its domain, and since $p \subseteq g$, $g(n)$ is defined.

Let $X=g^{-1}[1]$, the set of all "yes" members of the generic conditions. It is possible to give a name for X directly. Let $X = \{ (\check{n}, p) : p(n)=1 \}$, then val$(X, G) = X$. Now suppose $A \subseteq \omega$ in V. We claim that $X \neq A$. Let $D_A = \{ p : \exists n, n \in \mathrm{dom}(p)$ and $p(n)=1$ if and only if $n \in A \}$. D_A is dense. (Given any p, if n is not in p's domain, adjoin a value for n contrary to the status of "$n \in A$".) Then any $p \in G \cap D_A$ witnesses $X \neq A$. To summarize, X is a *new* subset of ω, necessarily infinite.

Replacing ω with $\omega \times \omega_2$, that is, consider instead finite partial functions whose inputs are of the form (n,α), with $n<\omega$ and $\alpha<\omega_2$, and whose outputs are 0 or 1, one gets ω_2 new subsets of ω. They are all distinct, by a density argument: given $\alpha<\beta<\omega_2$, let $D_{\alpha,\beta}=\{p:\exists n,\ p(n,\alpha)\neq p(n,\beta)\}$, then each $D_{\alpha,\beta}$ is dense, and a generic condition in it proves that the αth new set disagrees somewhere with the βth new set.

This is not yet the falsification of the continuum hypothesis. One must prove that no new maps have been introduced which map ω onto ω_1 or ω_1 onto ω_2. For example, if one considers instead Fin(ω,ω_1), finite partial functions from ω to ω_1, the first uncountable ordinal, one gets in $V[G]$ a bijection from ω to ω_1. In other words, ω_1 has *collapsed*, and in the forcing extension, is a countable ordinal.

The last step in showing the independence of the continuum hypothesis, then, is to show that Cohen forcing does not collapse cardinals. For this, a sufficient combinatorial property is that all of the antichains of this poset are countable.

The Countable Chain Condition

An antichain A of P is a subset such that if p and q are in A, then p and q are *incompatible* (written $p \perp q$), meaning there is no r in P such that $r \leq p$ and $r \leq q$. In the Borel sets example, incompatibility means $p\cap q$ has measure zero. In the finite partial functions example, incompatibility means that $p\cup q$ is not a function, in other words, p and q assign different values to some domain input.

P satisfies the countable chain condition (c.c.c.) if every antichain in P is countable. (The name, which is obviously inappropriate, is a holdover from older terminology. Some mathematicians write "c.a.c." for "countable antichain condition".)

It is easy to see that Bor(I) satisfies the c.c.c., because the measures add up to at most 1. Fin$(E,2)$ is also c.c.c., but the proof is more difficult.

Given an uncountable subfamily $W\subseteq$ Fin$(E,2)$, shrink W to an uncountable subfamily W_0 of sets of size n, for some $n<\omega$. If $p(e_1)=b_1$ for uncountably many $p\in W_0$, shrink to this uncountable subfamily W_1, and repeat, getting a finite set $\{(e_1,b_1), \ldots, (e_k,b_k)\}$, and an uncountable family W_k of incompatible conditions of size $n\in k$ such that every e is in at most countably many dom(p) for $p\in W_k$. Now pick an arbitrary $p\in W_k$, and pick from W_k any q that is not one of the countably many members that have a domain member in common with p. Then $p \cup \{ (e_1,b_1), \ldots, (e_k,b_k) \}$ and $q \cup \{ (e_1,b_1), \ldots, (e_k,b_k) \}$ are compatible, so W is not an antichain. In other words, Fin(E,2) antichains are countable.

The importance of antichains in forcing is that for most purposes, dense sets and maximal antichains are equivalent. A *maximal* antichain A is one that cannot be extended and still be an antichain. This means every element of $p \in$ P is compatible with some member of A. Their existence follows from Zorn's lemma. Given a maximal antichain A, let $D = \{ p : p \leq q$, some $q \in A \}$. D is dense, and $G \cap D \neq 0$ if and only if $G \cap A \neq 0$. Conversely, given a dense set D, Zorn's lemma shows there exists a maximal antichain $A \subseteq D$, and then $G \cap D \neq 0$ if and only if $G \cap A \neq 0$.

Assume P is c.c.c. Given $x,y \in V$, with $f{:}x \to y$ in $V[G]$, one can approximate f inside V as follows. Let u be a name for f (by the definition of $V[G]$) and let p be a condition which forces u to be a function from x to y. Define a function F whose domain is x by $F(a) = \{ b : \exists\ q \leq p,\ q$ forces $u(a\check{}) = b\check{}\ \}$. By definability of forcing, this definition makes sense within V. By coherence of forcing, different b's come from incompatible p's. By c.c.c., $F(a)$ is countable.

In summary, f is unknown in V, since it depends on G, but it is not wildly unknown for a c.c.c. forcing. One can identify a countable set of guesses for what the value of f is at any input, independent of G.

This has the following very important consequence. If in $V[G]$, $f{:}\alpha \to \beta$ is a surjection from one infinite ordinal to another, then there is a surjection $g{:}\omega \times \alpha \to \beta$ in V and consequently a surjection $h{:}\alpha \to \beta$ in V. In particular, cardinals cannot collapse. The conclusion is that $2^{\aleph_0} \geq \aleph_2$ in $V[G]$.

Easton Forcing

The exact value of the continuum in the above Cohen model, and variants like Fin($\omega \times \kappa$, 2) for cardinals κ in general, was worked out by Robert M. Solovay, who also worked out how to violate GCH (the generalized continuum hypothesis), for regular cardinals only, a finite number of times. For example, in the above Cohen model, if CH holds in V, then $2^{\aleph_0} = \aleph_2$ holds in $V[G]$.

W. B. Easton worked out the infinite and proper class version of violating the GCH for regular cardinals, basically showing the known restrictions (monotonicity, Cantor's theorem, and König's theorem) were the only ZFC provable restrictions.

Easton's work was notable in that it involved forcing with a proper class of conditions. In general, the method of forcing with a proper class of conditions will fail to give a model of ZFC. For example, Fin ($\omega \times$ On, 2), where "On" is the proper class of all ordinals, will make the continuum a proper class. Fin (ω, On) will introduce a countable enumeration of the ordinals. In both cases, the resulting $V[G]$ is visibly not a model of ZFC.

At one time, it was thought that more sophisticated forcing would also allow arbitrary variation in the powers of singular cardinals. But this has turned out to be a difficult, subtle and even surprising problem, with several more restrictions provable in ZFC, and with the forcing models depending on the consistency of various large cardinal properties. Many open problems remain.

Random Reals

In the Borel sets (Bor(I), $\subseteq$, I) example, the generic filter converges to a real number r, called a *random* real. A name for the decimal expansion of r (in the sense of the canonical set of decimal intervals that converge to r) can be given by letting $r = \{ (E^{\cdot}, E) : E = [k \cdot 10^{-n}, (k + 1)k \cdot 10^{-n}], 0 \leq k < 10^n \}$. This is, in some sense, just a subname of G.

To recover G from r, one takes those Borel subsets of I that "contain" r. Since the forcing poset is in V, but r is not in V, this containment is actually impossible. But there is a natural sense in which the interval [0.5, 0.6] in V "contains" a random real whose decimal expansion begins 0.5. This is formalized by the notion of "Borel code".

Every Borel set can, nonuniquely, be built up, starting from intervals with rational endpoints and applying the operations of complement and countable unions, a countable number of times. The record of such a construction is called a *Borel code*. Given a Borel set B in V, one recovers a Borel code, and then applies the same construction sequence in $V[G]$, getting a Borel set B^*. One can prove that one gets the same set independent of the construction of B, and that basic properties are preserved. For example, if $B \subseteq C$, then $B^* \subseteq C^*$. If B has measure zero, then B^* has measure zero.

So given r, a random real, one can show that $G = \{ B \text{ (in } V) : r \in B^* \text{ (in } V[G]) \}$. Because of the mutual interdefinability between r and G, one generally writes $V[r]$ for $V[G]$.

A different interpretation of reals in $V[G]$ was provided by Dana Scott. Rational numbers in $V[G]$ have names that correspond to countably many distinct rational values assigned to a maximal antichain of Borel sets, in other words, a certain rational-valued function on $I = [0,1]$. Real numbers in $V[G]$ then correspond to Dedekind cuts of such functions, that is, measurable functions.

Boolean-valued Models

Perhaps more clearly, the method can be explained in terms of Boolean-valued models. In these, any statement is assigned a truth value from some complete atomless Boolean algebra, rather than just a true/false value. Then an ultrafilter is picked in this Boolean algebra, which assigns values true/false to statements of our theory. The point is that the resulting theory has a model which contains this ultrafilter, which can be understood as a new model obtained by extending the old one with this ultrafilter. By picking a Boolean-valued model in an appropriate way, we can get a model that has the desired property. In it, only statements which must be true (are "forced" to be true) will be true, in a sense (since it has this extension/minimality property).

Meta-mathematical Explanation

In forcing we usually seek to show some sentence is consistent with ZFC (or optionally some extension of ZFC). One way to interpret the argument is that we assume ZFC is consistent and use it to prove ZFC combined with our new sentence is also consistent.

Each "condition" is a finite piece of information – the idea is that only finite pieces are relevant for consistency, since by the compactness theorem a theory is satisfiable if and only if every finite subset of its axioms is satisfiable. Then, we can pick an infinite set of consistent conditions to extend our model. Thus, assuming consistency of set theory, we prove consistency of the theory extended with this infinite set.

Logical Explanation

By Gödel's incompleteness theorem one cannot prove the consistency of any sufficiently strong formal theory, such as ZFC, using only the axioms of the theory itself, unless the theory itself is inconsistent. Consequently mathematicians do not attempt to prove the consistency of ZFC using only the axioms of ZFC, or to prove ZFC+H is consistent for any hypothesis H using only ZFC+H. For this reason the aim of a consistency proof is to prove the consistency of ZFC + H relative to consistency of ZFC. Such problems are known as problems of relative consistency. In fact one proves

$$(*)\quad ZFC \vdash \text{Con}(ZFC) \to \text{Con}(ZFC + H).$$

We will give the general schema of relative consistency proofs. Because any proof is finite it uses finite number of axioms.

$$ZFC + \neg\text{Con}(ZFC + H) \vdash \exists T(\text{Fin}(T) \wedge T \subset ZFC \wedge (T \vdash \neg H)).$$

For any given proof ZFC can verify validity of this proof. This is provable by induction by the length of the proof.

$$ZFC \vdash \forall T((T \vdash \neg H) \to (ZFC \vdash (T \vdash \neg H))).$$

Now we obtain

$$ZFC + \neg\text{Con}(ZFC + H) \vdash \exists T(\text{Fin}(T) \wedge T \subset ZFC \wedge (ZFC \vdash (T \vdash \neg H))).$$

If we prove following

$$(**) ZFC \vdash \forall T(\text{Fin}(T) \wedge T \subset ZFC \to (ZFC \vdash \text{Con}(T + H)))$$

we can conclude that

$$ZFC + \neg\text{Con}(ZFC + H) \vdash \exists T(\text{Fin}(T) \wedge T \subset ZFC \wedge (ZFC \vdash (T \vdash \neg H)) \wedge (ZFC \vdash \text{Con}(T + H)))$$

which is equivalent to

$$ZFC + \neg\text{Con}(ZFC + H) \vdash \neg\text{Con}(ZFC)$$

which gives (*). Core of relative consistency proof is proving of (**). One have to construct ZFC proof of Con($T + H$) for any given finite set T of ZFC axioms (by ZFC instruments of course). (No universal proof of Con($T + H$) of course.)

In ZFC is provable that for any condition p the set of formulas (evaluated by names) forced by p is deductive closed. Also, for any ZFC axiom ZFC proves that this axiom is forced by 1. Then proving that there is at least one condition which forces H suffices.

In case of Boolean valued forcing procedure is similar – one have to prove that Boolean value of H is not 0.

Another approach is using reflection theorem. For any given finite set of ZFC axioms there is ZFC proof that this set of axioms has countable transitive model. For any given finite set T of ZFC axioms there is finite set T' of ZFC axioms such that ZFC proves that if countable transitive model M satisfies T' then $M[G]$ satisfies T. One have to prove that there is finite set T'' of ZFC axioms such that if countable transitive model M satisfies T'' then $M[G]$ satisfies considered hypothesis H. Then for any given finite set T of ZFC axioms ZFC proves $\mathrm{Con}(T + H)$.

Sometimes in (**) some stronger theory S than ZFC is used for proving $\mathrm{Con}(T + H)$. Then we have proof of consistency of ZFC + H relative to consistency of S. Note that $ZFC \vdash \mathrm{Con}(ZFC) \leftrightarrow \mathrm{Con}(ZFL)$, where ZFL is ZF + V = L (axiom of constructibility).

Cardinal Function

In mathematics, a cardinal function (or cardinal invariant) is a function that returns cardinal numbers.

Cardinal Functions in Set Theory

- The most frequently used cardinal function is a function which assigns to a set "A" its cardinality, denoted by $|A|$.
- Aleph numbers and beth numbers can both be seen as cardinal functions defined on ordinal numbers.
- Cardinal arithmetic operations are examples of functions from cardinal numbers (or pairs of them) to cardinal numbers.
- Cardinal characteristics of a (proper) ideal I of subsets of X are:

$$\text{add}(I) = \min\{|\mathcal{A}| : \mathcal{A} \subseteq I \wedge \bigcup \mathcal{A} \notin I\}.$$

The "additivity" of I is the smallest number of sets from I whose union is not in I any more. As any ideal is closed under finite unions, this number is always at least $\aleph_0$; if I is a σ-ideal, then $\text{add}(I) \geq \aleph_1$.

$$\text{cov}(I) = \min\{|\mathcal{A}| : \mathcal{A} \subseteq I \wedge \bigcup \mathcal{A} = X\}.$$

The "covering number" of I is the smallest number of sets from I whose union is all of X. As X itself is not in I, we must have $\text{add}(I) \leq \text{cov}(I)$.

$$\text{non}(I) = \min\{|A| : A \subseteq X \wedge A \notin I\},$$

The "uniformity number" of I (sometimes also written $\text{unif}(I)$) is the size of the smallest set not in I. Assuming I contains all singletons, $\text{add}(I) \leq \text{non}(I)$.

$$\text{cof}(I) = \min\{|\mathcal{B}| : \mathcal{B} \subseteq I \wedge (\forall A \in I)(\exists B \in \mathcal{B})(A \subseteq B)\}.$$

The "cofinality" of I is the cofinality of the partial order $(I, \subseteq)$. It is easy to see that we must have $\text{non}(I) \leq \text{cof}(I)$ and $\text{cov}(I) \leq \text{cof}(I)$.

In the case that I is an ideal closely related to the structure of the reals, such as the ideal of Lebesgue null sets or the ideal of meagre sets, these cardinal invariants are referred to as cardinal characteristics of the continuum.

- For a preordered set $(\mathbb{P}, \sqsubseteq)$ the bounding number $\mathfrak{b}(\mathbb{P})$ and dominating number $\mathfrak{d}(\mathbb{P})$ is defined as

$$\mathfrak{b}(\mathbb{P}) = \min\{|Y| : Y \subseteq \mathbb{P} \wedge (\forall x \in \mathbb{P})(\exists y \in Y)(y \not\sqsubseteq x)\},$$

$$\mathfrak{d}(\mathbb{P}) = \min\{|Y| : Y \subseteq \mathbb{P} \wedge (\forall x \in \mathbb{P})(\exists y \in Y)(x \sqsubseteq y)\}$$

- In PCF theory the cardinal function $pp_\kappa(\lambda)$ is used.

Cardinal Functions in Algebra

Examples of cardinal functions in algebra are:

- Index of a subgroup H of G is the number of cosets.
- Dimension of a vector space V over a field K is the cardinality of any Hamel basis of V.

- More generally, for a free module M over a ring R we define rank rank(M) as the cardinality of any basis of this module.
- For a linear subspace W of a vector space V we define codimension of W (with respect to V).
- For any algebraic structure it is possible to consider the minimal cardinality of generators of the structure.
- For algebraic extensions algebraic degree and separable degree are often employed (note that the algebraic degree equals the dimension of the extension as a vector space over the smaller field).

For non-algebraic field extensions transcendence degree is likewise used.

Set-theoretic Topology

In mathematics, set-theoretic topology is a subject that combines set theory and general topology. It focuses on topological questions that are independent of Zermelo–Fraenkel set theory(ZFC). A famous problem is the normal Moore space question, a question in general topology that was the subject of intense research. The answer to the normal Moore space question was eventually proved to be independent of ZFC.

Objections to Set Theory as a Foundation for Mathematics

From set theory's inception, some mathematicians have objected to it as a foundation for mathematics. The most common objection to set theory, one Kronecker voiced in set theory's earliest years, starts from the constructivist view that mathematics is loosely related to computation. If this view is granted, then the treatment of infinite sets, both in naive and in axiomatic set theory, introduces into mathematics methods and objects that are not computable even in principle. Ludwig Wittgenstein questioned the way Zermelo–Fraenkel set theory handled infinities. Wittgenstein's views about the foundations of mathematics were later criticised by Georg Kreisel and Paul Bernays, and investigated by Crispin Wright, among others.

Category theorists have proposed topos theory as an alternative to traditional axiomatic set theory. Topos theory can interpret various alternatives to that theory, such as constructivism, finite set theory, and computable set theory.

Boolean Algebra

In mathematics and mathematical logic, Boolean algebra is the subarea of algebra in which the values of the variables are the truth values *true* and *false*, usually denoted 1 and 0 respectively. Instead of elementary algebra where the values of the variables are numbers, and the main operations are addition and multiplication, the main operations of Boolean algebra are the conjunction *and*, denoted $\wedge$, the disjunction *or*, denoted $\vee$, and the negation *not*, denoted $\neg$.

Boolean algebra was introduced in 1854 by George Boole in his book *An Investigation of the Laws of Thought*. According to Huntington the term "Boolean algebra" was first suggested by Sheffer in 1913.

Boolean algebra has been fundamental in the development of computer science and is yet the basis of the abstract description of digital circuits. It is also used in digital logic, computer programming, set theory, and statistics.

History

Boole's algebra predated the modern developments in abstract algebra and mathematical logic; it is however seen as connected to the origins of both fields. In an abstract setting, Boolean algebra was perfected in the late 19th century by Jevons, Schröder, Huntington, and others until it reached the modern conception of an (abstract) mathematical structure. For example, the empirical observation that one can manipulate expressions in the algebra of sets by translating them into expressions in Boole's algebra is explained in modern terms by saying that the algebra of sets is *a* Boolean algebra (note the indefinite article). In fact, M. H. Stone proved in 1936 that every Boolean algebra is isomorphic to a field of sets.

In the 1930s, while studying switching circuits, Claude Shannon observed that one could also apply the rules of Boole's algebra in this setting, and he introduced switching algebra as a way to analyze and design circuits by algebraic means in terms of logic gates. Shannon already had at his disposal the abstract mathematical apparatus, thus he cast his switching algebra as the two-element Boolean algebra. In circuit engineering settings today, there is little need to consider other Boolean algebras, thus "switching algebra" and "Boolean algebra" are often used interchangeably. Efficient implementation of Boolean functions is a fundamental problem in the design of combinatorial logic circuits. Modern electronic design automation tools for VLSI circuits often rely on an efficient representation of Boolean functions known as (reduced ordered) binary decision diagrams (BDD) for logic synthesis and formal verification.

Logic sentences that can be expressed in classical propositional calculus have an equivalent expression in Boolean algebra. Thus, Boolean logic is sometimes used to denote propositional calculus performed in this way. Boolean algebra is not sufficient to capture logic formulas using quantifiers, like those from first order logic. Although the development of mathematical logic did not follow Boole's programme, the connection between his algebra and logic was later put on firm ground in the setting of algebraic logic, which also studies the algebraic systems of many other logics. The problem of determining whether the variables of a given Boolean (propositional) formula can be assigned in such a way as to make the formula evaluate to true is called the Boolean satisfiability problem (SAT), and is of importance to theoretical computer science, being the first problem shown to be NP-complete. The closely related model of computation known as a Boolean circuit relates time complexity (of an algorithm) to circuit complexity.

Variables

Whereas, in elementary algebra, the variables represent mainly numbers, in Boolean algebra, the variables represent the truth values *false* and *true*, which are identified with the bits (or binary digits) and denoted by 0 and 1. They should not be confused with the integers 0 and 1, but may also identified with the elements of the field with two elements.

Boolean algebra also deals with functions which have their values in the set {0, 1}. A sequence of bits is a commonly used such function. Another common example is the subsets of a set E: to a subset F of E is associated the indicator function that takes the value 1 on E and 0 outside E.

Like in usual algebra, most of the theory may be developed without considering explicit values for the variables.

Completeness

At this point we can now claim to have defined Boolean algebra, in the sense that the laws we have listed up to now entail the rest of the subject. The laws *Complementation* 1 and 2, together with the monotone laws, suffice for this purpose and can therefore be taken as one possible *complete* set of laws or axiomatization of Boolean algebra. Every law of Boolean algebra follows logically from these axioms. Furthermore Boolean algebras can then be defined as the models of these axioms as treated in the section thereon.

To clarify, writing down further laws of Boolean algebra cannot give rise to any new consequences of these axioms, nor can it rule out any model of them. Had we stopped listing laws too soon, there would have been Boolean laws that did not follow from those on our list, and moreover there would have been models of the listed laws that were not Boolean algebras.

This axiomatization is by no means the only one, or even necessarily the most natural given that we did not pay attention to whether some of the axioms followed from others but simply chose to stop when we noticed we had enough laws, treated further in the section on axiomatizations. Or the intermediate notion of axiom can be sidestepped altogether by defining a Boolean law directly as any tautology, understood as an equation that holds for all values of its variables over 0 and 1. All these definitions of Boolean algebra can be shown to be equivalent.

Boolean algebra has the interesting property that $x = y$ can be proved from any non-tautology. This is because the substitution instance of any non-tautology obtained by instantiating its variables with constants 0 or 1 so as to witness its non-tautologyhood reduces by equational reasoning to $0 = 1$. For example the non-tautologyhood of $x \wedge y = x$ is witnessed by $x = 1$ and $y = 0$ and so taking this as an axiom would allow us to infer $1 \wedge 0 = 1$ as a substitution instance of the axiom and hence $0 = 1$. We can then show $x = y$ by the reasoning $x = x \wedge 1 = x \wedge 0 = 0 = 1 = y \vee 1 = y \vee 0 = y$.

Duality Principle

There is nothing magical about the choice of symbols for the values of Boolean algebra. We could rename 0 and 1 to say α and β, and as long as we did so consistently throughout it would still be Boolean algebra, albeit with some obvious cosmetic differences.

But suppose we rename 0 and 1 to 1 and 0 respectively. Then it would still be Boolean algebra, and moreover operating on the same values. However it would not be identical to our original Boolean algebra because now we find $\vee$ behaving the way $\wedge$ used to do and vice versa. So there are still some cosmetic differences to show that we've been fiddling with the notation, despite the fact that we're still using 0s and 1s.

But if in addition to interchanging the names of the values we also interchange the names of the two binary operations, *now* there is no trace of what we have done. The end product is completely indistinguishable from what we started with. We might notice that the columns for $x \wedge y$ and $x \vee y$ in the truth tables had changed places, but that switch is immaterial.

When values and operations can be paired up in a way that leaves everything important unchanged when all pairs are switched simultaneously, we call the members of each pair dual to each other. Thus 0 and 1 are dual, and $\wedge$ and $\vee$ are dual. The Duality Principle, also called De Morgan duality, asserts that Boolean algebra is unchanged when all dual pairs are interchanged.

One change we did not need to make as part of this interchange was to complement. We say that complement is a self-dual operation. The identity or do-nothing operation x (copy the input to the output) is also self-dual. A more complicated example of a self-dual operation is $(x \wedge y) \vee (y \wedge z) \vee (z \wedge x)$. It can be shown that self-dual operations must take an odd number of arguments; thus there can be no self-dual binary operation.

The principle of duality can be explained from a group theory perspective by fact that there are exactly four functions that are one-to-one mappings (automorphisms) of the set of Boolean polynomials back to itself: the identity function, the complement function, the dual function and the contradual function (complemented dual). These four functions form a group under function composition, isomorphic to the Klein four-group, acting on the set of Boolean polynomials.

Diagrammatic Representations

Venn Diagrams: A Venn diagram is a representation of a Boolean operation using shaded overlapping regions. There is one region for each variable, all circular in the examples here. The interior and exterior of region x corresponds respectively to the values 1 (true) and 0 (false) for variable x. The shading indicates the value of the operation for each combination of regions, with dark denoting 1 and light 0 (some authors use the opposite convention).

The three Venn diagrams in the figure below represent respectively conjunction $x \wedge y$, disjunction $x \vee y$, and complement $\neg x$.

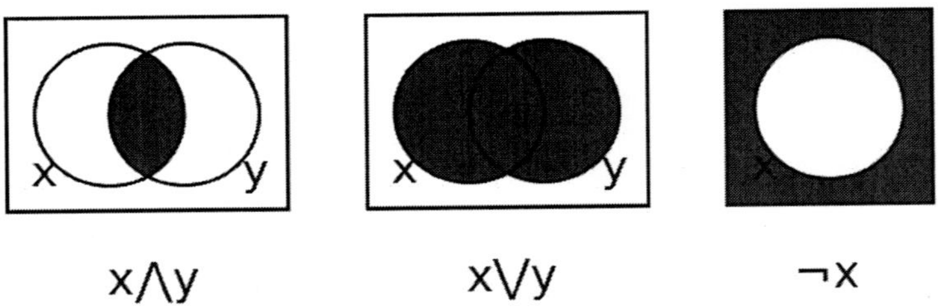

Figure: *Venn diagrams for conjunction, disjunction, and complement*

For conjunction, the region inside both circles is shaded to indicate that $x \wedge y$ is 1 when both variables are 1. The other regions are left unshaded to indicate that $x \wedge y$ is 0 for the other three combinations. The second diagram represents disjunction $x \wedge y$ by shading those regions that lie inside either or both circles. The third diagram represents complement $\neg x$ by shading the region *not* inside the circle.

While we have not shown the Venn diagrams for the constants 0 and 1, they are trivial, being respectively a white box and a dark box, neither one containing a circle. However we could put a circle for x in those boxes, in which case each would denote a function of one argument, x, which returns the same value independently of x, called a constant function. As far as their outputs are concerned, constants and constant functions are indistinguishable; the difference is that a constant takes no arguments, called a *zeroary* or *nullary* operation, while a constant function takes one argument, which it ignores, and is a *unary* operation.

Venn diagrams are helpful in visualizing laws. The commutativity laws for $\wedge$ and $\vee$ can be seen from the symmetry of the diagrams: a binary operation that was not commutative would not have a symmetric diagram because interchanging x and y would have the effect of reflecting the diagram horizontally and any failure of commutativity would then appear as a failure of symmetry.

Idempotence of $\wedge$ and $\vee$ can be visualized by sliding the two circles together and noting that the shaded area then becomes the whole circle, for both $\wedge$ and $\vee$.

To see the first absorption law, $x \wedge (x \vee y) = x$, start with the diagram in the middle for $x \wedge y$ and note that the portion of the shaded area in common with the x circle is the whole of the x circle. For the second absorption law, $x \vee (x \wedge y) = x$, start with the left diagram for $x \wedge y$ and note that shading the whole of the x circle results in just the x circle being shaded, since the previous shading was inside the x circle.

The double negation law can be seen by complementing the shading in the third diagram for $\neg x$, which shades the x circle.

To visualize the first De Morgan's law, $(\neg x) \wedge (\neg y) = \neg(x \vee y)$, start with the middle diagram for $x \vee y$ and complement its shading so that only the region outside both circles is shaded, which is what the right hand side of the law describes. The result is the same as if we shaded that region which is both outside the x circle *and* outside the y circle, i.e. the conjunction of their exteriors, which is what the left hand side of the law describes.

The second De Morgan's law, $(\neg x) \vee (\neg y) = \neg(x \wedge y)$, works the same way with the two diagrams interchanged.

The first complement law, $x \wedge \neg x = 0$, says that the interior and exterior of the x circle have no overlap. The second complement law, $x \vee \neg x = 1$, says that everything is either inside or outside the x circle.

Digital Logic Gates

Digital logic is the application of the Boolean algebra of 0 and 1 to electronic hardware consisting of logic gates connected to form a circuit diagram. Each gate implements a Boolean operation, and is depicted schematically by a shape indicating the operation. The shapes associated with the gates for conjunction (AND-gates), disjunction (OR-gates), and complement (inverters) are as follows.

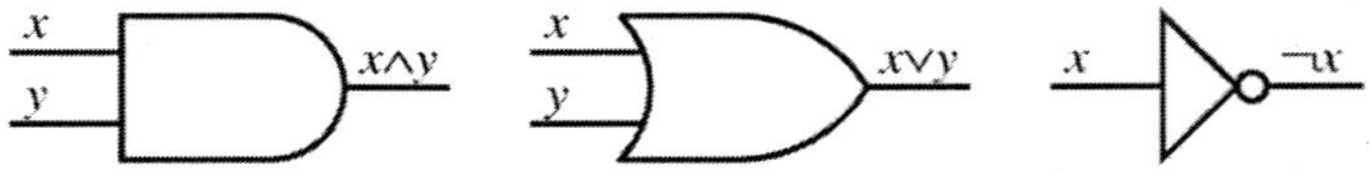

Figure: *The lines on the left of each gate represent input wires or* ports. *The value of the input is represented by a voltage on the lead. For so-called "active-high" logic 0 is represented by a voltage close to zero or "ground" while 1 is represented by a voltage close to the supply voltage; active-low reverses this. The line on the right of each gate represents the output port, which normally follows the same voltage conventions as the input ports.*

Complement is implemented with an inverter gate. The triangle denotes the operation that simply copies the input to the output; the small circle on the output denotes the actual inversion complementing the input. The convention of putting such a circle on any port means that the signal passing through this port is complemented on the way through, whether it is an input or output port.

There being eight ways of labelling the three ports of an AND-gate or OR-gate with inverters, this convention gives a wide range of possible Boolean operations realized as such gates so decorated. Not all combinations are distinct however: any labelling of AND-gate ports with inverters realizes the same Boolean operation as the opposite labelling of OR-gate ports (a given port of the AND-gate is labelled with an inverter if and only if the corresponding port of the OR-gate is not so labelled). This follows from De Morgan's laws.

If we complement all ports on every gate, and interchange AND-gates and OR-gates, as in Figure 4 below, we end up with the same operations as we started with, illustrating both De Morgan's laws and the Duality Principle. Note that we did not need to change the triangle part of the inverter, illustrating self-duality for complement.

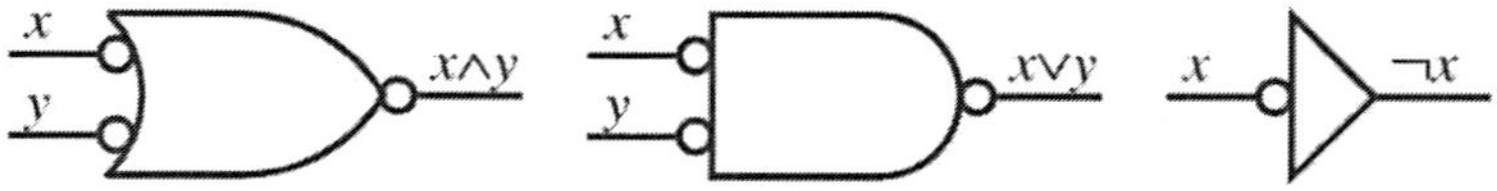

Because of the pairwise identification of gates via the Duality Principle, even though 16 schematic symbols can be manufactured from the two basic binary gates AND and OR by furnishing their ports with inverters (circles), they only represent eight Boolean operations, namely those operations with an odd number of ones in their truth table. Altogether there are 16 binary Boolean operations, the other eight being those with an even number of ones in their truth table, namely the following. The constant 0, viewed as a binary operation that ignores both its inputs, has no ones, the six operations x, y, $\neg x$, $\neg y$ (as binary operations that ignore one input), $x \oplus y$, and $x \equiv y$ have two ones, and the constant 1 has four ones.

Concrete Boolean Algebras

A concrete Boolean algebra or field of sets is any nonempty set of subsets of a given set X closed under the set operations of union, intersection, and complement relative to X.

(As an aside, historically X itself was required to be nonempty as well to exclude the degenerate or one-element Boolean algebra, which is the one exception to the rule that all Boolean algebras satisfy the same equations since the degenerate algebra satisfies every equation. However this exclusion conflicts with the preferred purely equational definition of "Boolean algebra," there being no way to rule out the one-element algebra using only equations— $0 \neq 1$ does not count, being a negated equation. Hence modern authors allow the degenerate Boolean algebra and let X be empty.)

Example 1: The power set 2^X of X, consisting of all subsets of X. Here X may be any set: empty, finite, infinite, or even uncountable.

Example 2: The empty set and X. This two-element algebra shows that a concrete Boolean algebra can be finite even when it consists of subsets of an infinite set. It can be seen that every field of subsets of X must contain the empty set and X. Hence no smaller example is possible, other than the degenerate algebra obtained by taking X to be empty so as to make the empty set and X coincide.

Example 3: The set of finite and cofinite sets of integers, where a cofinite set is one omitting only finitely many integers. This is clearly closed under complement, and is closed under union because the union of a cofinite set with any set is cofinite, while the union of two finite sets is finite. Intersection behaves like union with "finite" and "cofinite" interchanged.

Example 4: For a less trivial example of the point made by Example 2, consider a Venn diagram formed by n closed curves partitioning the diagram into 2^n regions, and let X be the (infinite) set of all points in the plane not on any curve but somewhere within the diagram. The interior of each region is thus an infinite subset of X, and every point in X is in exactly one region. Then the set of all 2^{2n} possible unions of regions (including the empty set obtained as the union of the empty set of regions and X obtained as the union of all 2^n regions) is closed under union, intersection, and complement relative to X and therefore forms a concrete Boolean algebra. Again we have finitely many subsets of an infinite set forming a concrete Boolean algebra, with Example 2 arising as the case $n = 0$ of no curves.

Subsets as bit Vectors

A subset Y of X can be identified with an indexed family of bits with index set X, with the bit indexed by $x \in X$ being 1 or 0 according to whether or not $x \in Y$. (This is the so-called characteristic function notion of a subset.) For example a 32-bit computer word consists of 32 bits indexed by the set $\{0,1,2,\ldots,31\}$, with 0 and 31 indexing the low and high order bits respectively. For a smaller example, if $X = \{a,b,c\}$ where a, b, c are viewed as bit positions in that order from left to right, the eight subsets, $\{c\}$, $\{b\}$, $\{b,c\}$, $\{a\}$, $\{a,c\}$, $\{a,b\}$, and $\{a,b,c\}$ of X can be identified with the respective bit vectors 000, 001, 010, 011, 100, 101, 110, and 111. Bit vectors indexed by the set of natural numbers are infinite sequences of bits, while those indexed by the reals in the unit interval [0,1] are packed too densely to be able to write conventionally but nonetheless form well-defined indexed families (imagine colouring every point of the interval [0,1] either black or white independently; the black points then form an arbitrary subset of [0,1]).

From this bit vector viewpoint, a concrete Boolean algebra can be defined equivalently as a nonempty set of bit vectors all of the same length (more generally, indexed by the same set) and closed under the bit vector operations of bitwise $\wedge$, $\vee$, and ¬, as in 1010 ∧ 0110 = 0010, 1010 ∨ 0110 = 1110, and ¬1010 = 0101, the bit vector realizations of intersection, union, and complement respectively.

Two-element Boolean Algebra

In mathematics and abstract algebra, the two-element Boolean algebra is the Boolean algebra whose *underlying set* (or universe or *carrier*) B is the Boolean domain. The elements of the Boolean domain are 1 and 0 by convention, so that $B = \{0, 1\}$. Paul Halmos's name for this algebra "2" has some following in the literature, and will be employed here.

Definition

B is a partially ordered set and the elements of B are also its bounds.

An operation of arity n is a mapping from B^n to B. Boolean algebra consists of two binary operations and unary complementation. The binary operations have been named and notated in various ways. Here they are called 'sum' and 'product', and notated by infix '+' and '.', respectively. Sum and product commute and associate, as in the usual algebra of real numbers. As for the order of operations, brackets are decisive if present. Otherwise '.' precedes '+'. Hence $A.B + C$ is parsed as $(A.B) + C$ and not as $A.(B + C)$. Complementation is denoted by writing an overbar over its argument. The numerical analog of the complement of X is $1 - X$. In the language of universal algebra, a Boolean algebra is a $\langle B,+,.,\bar{.},1,0\rangle$ algebra of type $\langle 2,2,1,0,0\rangle$.

Either one-to-one correspondence between {0,1} and {*True*,*False*} yields classical bivalent logic in equational form, with complementation read as NOT. If 1 is read as *True*, '+' is read as OR, and '.' as AND, and vice versa if 1 is read as *False*.

Some Basic Identities

2 can be seen as grounded in the following trivial "Boolean" arithmetic:

- 1+1 = 1+0 = 0+1 = 1
- 0.0 = 0.1 = 1.0 = 0
- 1.1 = 1

- $0+0 = 0$
- $\overline{1} = 0$
- $\overline{0} = 1.$

Note that:

- '+' and '.' work exactly as in numerical arithmetic, except that 1+1=1.'+' and '.' are derived by analogy from numerical arithmetic; simply set any nonzero number to 1.
- Swapping 0 and 1, and '+' and '.' preserves truth; this is the essence of the duality pervading all Boolean algebras.

This Boolean arithmetic suffices to verify any equation of 2, including the axioms, by examining every possible assignment of 0s and 1s to each variable.

The following equations may now be verified:

- $A + A = A$
- $A.A = A$
- $A + 0 = A$
- $A + 1 = 1$
- $A.0 = 0$
- $\overline{\overline{A}} = A.$

Each of '+' and '.' distributes over the other:

- $A.(B + C) = A.B + A.C;$
- $A + (B.C) = (A + B).(A + C).$

That '.' distributes over '+' agrees with elementary algebra, but not '+' over '.'. For this and other reasons, a sum of products (leading to a NAND synthesis) is more commonly employed than a product of sums (leading to a NOR synthesis).

Each of '+' and '.' can be defined in terms of the other and complementation:

- $A.B = \overline{\overline{A} + \overline{B}}$
- $A + B = \overline{\overline{A}.\overline{B}}.$

We only need one binary operation, and concatenation suffices to denote it. Hence concatenation and overbar suffice to notate 2. This notation is also that of Quine's Boolean term schemata. Letting (*X*) denote the complement of *X* and "()" denote either 0 or 1 yields the syntax of the primary algebra.

A *basis* for 2 is a set of equations, called axioms, from which all of the above equations (and more) can be derived. There are many known bases for all Boolean algebras and hence for 2. An elegant basis notated using only concatenation and overbar is:

1. $ABC = BCA$ (Concatenation commutes, associates)
2. $\bar{A}A = 1$ (2 is a complemented lattice, with an upper bound of 1)
3. $A0 = A$ (0 is the lower bound).
4. $A\overline{AB} = A\bar{B}$ (2 is a distributive lattice)

If 0=1, (1)-(3) are the axioms for an abelian group.

(1) only serves to prove that concatenation commutes and associates. First assume that (1) associates from either the left or the right, then prove commutativity. Then prove association from the other direction. Associativity is simply association from the left and right combined.

This basis makes for an easy approach to proof, called calculation, that proceeds by simplifying expressions to 0 or 1, by invoking axioms (2)–(4), and the elementary identities $AA = A, \bar{\bar{A}} = A, 1 + A = 1$, and the distributive law.

Metatheory

De Morgan's theorem states that if one does the following, in the given order, to any Boolean function:

- Complement every variable;
- Swap + and . operators (taking care to add brackets to ensure the order of operations remains the same);
- Complement the result, the result is logically equivalent to what you started with. Repeated application of De Morgan's theorem to parts of a function can be used to drive all complements down to the individual variables.

A powerful and nontrivial metatheorem states that any theorem of 2 holds for all Boolean algebras. Conversely, an identity that holds for an arbitrary nontrivial Boolean algebra also holds in 2. Hence all the mathematical content of Boolean algebra is captured by 2. This theorem is useful because any equation in 2 can be verified by a decision procedure. Logicians refer to this fact as "2 is decidable". All known decision procedures require a number of steps that is an exponential function of the number of variables N appearing in the equation to be verified. Whether there exists a decision procedure whose steps are a polynomial function of N falls under the P = NP conjecture.

Boolean Algebras: the Definition

The Boolean algebras we have seen so far have all been concrete, consisting of bit vectors or equivalently of subsets of some set. Such a Boolean algebra consists of a set and operations on that set which can be *shown* to satisfy the laws of Boolean algebra. Instead of showing that the Boolean laws are satisfied, we can instead postulate a set X, two binary operations on X, and one unary operation, and *require* that those operations satisfy the laws of Boolean algebra. The elements of X need not be bit vectors or subsets but can be anything at all. This leads to the more general *abstract* definition.

A Boolean algebra is any set with binary operations $\wedge$ and $\vee$ and a unary operation $\neg$ thereon satisfying the Boolean laws.

For the purposes of this definition it is irrelevant how the operations came to satisfy the laws, whether by fiat or proof. All concrete Boolean algebras satisfy the laws (by proof rather than fiat), whence every concrete Boolean algebra is a Boolean algebra according to our definitions. This axiomatic definition of a Boolean algebra as a set and certain operations satisfying certain laws or axioms is entirely analogous to the abstract definitions of group, ring, field etc. characteristic of modern or abstract algebra.

Given any complete axiomatization of Boolean algebra, such as the axioms for a complemented distributive lattice, a sufficient condition for an algebraic structure of this kind to satisfy all the Boolean laws is that it satisfy just those axioms. The following is therefore an equivalent definition.

A Boolean algebra is a complemented distributive lattice.

The section on axiomatization lists other axiomatizations, any of which can be made the basis of an equivalent definition.

Representable Boolean Algebras

Although every concrete Boolean algebra is a Boolean algebra, not every Boolean algebra need be concrete. Let n be a square-free positive integer, one not divisible by the square of an integer, for example 30 but not 12. The operations of greatest common divisor, least common multiple, and division into n (that is, $\neg x = n/x$), can be shown to satisfy all the Boolean laws when their arguments range over the positive divisors of n. Hence those divisors form a Boolean algebra. These divisors are not subsets of a set, making the divisors of n a Boolean algebra that is not concrete according to our definitions.

However if we *represent* each divisor of n by the set of its prime factors, we find that this nonconcrete Boolean algebra is isomorphic to the concrete Boolean algebra consisting of all sets of prime factors of n, with union corresponding to least common multiple, intersection to greatest common divisor, and complement to division into n. So this example while not technically concrete is at least "morally" concrete via this representation, called an isomorphism. This example is an instance of the following notion.

A Boolean algebra is called representable when it is isomorphic to a concrete Boolean algebra.

The obvious next question is answered positively as follows.

Every Boolean algebra is representable.

That is, up to isomorphism, abstract and concrete Boolean algebras are the same thing. This quite nontrivial result depends on the Boolean prime ideal theorem, a choice principle slightly weaker than the axiom of choice, and is treated in more detail in the article Stone's representation theorem for Boolean algebras. This strong relationship implies a weaker result strengthening the observation in the previous subsection to the following easy consequence of representability.

The laws satisfied by all Boolean algebras coincide with those satisfied by the prototypical Boolean algebra.

It is weaker in the sense that it does not of itself imply representability. Boolean algebras are special here, for example a relation algebra is a Boolean algebra with additional structure but it is not the case that every relation algebra is representable in the sense appropriate to relation algebras.

Axiomatizing Boolean Algebra

The above definition of an abstract Boolean algebra as a set and operations satisfying "the" Boolean laws raises the question, what are those laws? A simple-minded answer is "all Boolean laws," which can be defined as all equations that hold for the Boolean algebra of 0 and 1. Since there are infinitely many such laws this is not a terribly satisfactory answer in practice, leading to the next question: does it suffice to require only finitely many laws to hold?

In the case of Boolean algebras the answer is yes. In particular the finitely many equations we have listed above suffice. We say that Boolean algebra is finitely axiomatizable or finitely based.

Can this list be made shorter yet? Again the answer is yes. To begin with, some of the above laws are implied by some of the others. A sufficient subset of the above laws consists of the pairs of associativity, commutativity, and absorption laws, distributivity of $\wedge$ over $\vee$ (or the other distributivity law—one suffices), and the two complement laws. In fact this is the traditional axiomatization of Boolean algebra as a complemented distributive lattice.

By introducing additional laws not listed above it becomes possible to shorten the list yet further. In 1933 Edward Huntington showed that if the basic operations are taken to be $x \wedge y$ and $\neg x$, with $x \wedge y$ considered a derived operation (e.g. via De Morgan's law in the form $x \wedge y = \neg(\neg x \wedge \neg y)$), then the equation $\neg(\neg x \wedge \neg y) \wedge \neg(\neg x \wedge y) = x$ along with the two equations expressing associativity and commutativity of $\wedge$ completely axiomatized Boolean algebra. When the only basic operation is the binary NAND operation $\neg(x \wedge y)$, Stephen Wolfram has proposed in his book *A New Kind of Science* the single axiom $(((xy)z)(x((xz)x))) = z$ as a one-equation axiomatization of Boolean algebra, where for convenience here xy denotes the NAND rather than the AND of x and y.

Applications

Two-valued Logic: Boolean algebra as the calculus of two values is fundamental to digital logic, computer programming, and mathematical logic, and is also used in other areas of mathematics such as set theory and statistics.

Digital logic codes its symbols in various ways: as voltages on wires in high-speed circuits and capacitive storage devices, as orientations of a magnetic domain in ferromagnetic storage devices, as holes in punched cards or paper tape, and so on. Now it is possible to code more than two symbols in any given medium. For example one might use respectively 0, 1, 2, and 3 volts to code a four-symbol alphabet on a wire, or holes of different sizes in a punched card. In practice however the tight constraints of high speed, small size, and low power combine to make noise a major factor. This makes it hard to distinguish between symbols when there are many of them at a single site. Rather than attempting to distinguish between four voltages on one wire, digital designers have settled on two voltages per wire, high and low. To obtain four symbols one uses two wires, and so on.

Programmers programming in machine code, assembly language, and other programming languages that expose the low-level digital structure of the data registers operate on whatever symbols were chosen for the hardware, invariably bit vectors in modern computers for the above reasons. Such languages support both the numeric operations of addition, multiplication, etc. performed on words interpreted as integers, as well as the logical operations of disjunction, conjunction, etc. performed bit-wise on words interpreted as bit vectors. Programmers therefore have the option of working in and applying the laws of either numeric algebra or Boolean algebra as needed. A core differentiating feature is carry propagation with the former but not the latter.

Other areas where two values is a good choice are the law and mathematics. In everyday relaxed conversation, nuanced or complex answers such as "maybe" or "only on the weekend" are acceptable. In more focused situations such as a court of law or theorem-based mathematics however it is deemed advantageous to frame questions so as to admit a simple yes-or-no answer—is the defendant guilty or not guilty, is the proposition true or false—and to disallow any other answer. However much of a straitjacket this might prove in practice for the respondent, the principle of the simple yes-no question has become a central feature of both judicial and mathematical logic, making two-valued logic deserving of organization and study in its own right.

A central concept of set theory is membership. Now an organization may permit multiple degrees of membership, such as novice, associate, and full. With sets however an element is either in or out. The candidates for membership in a set work just like the wires in a digital computer: each candidate is either a member or a nonmember, just as each wire is either high or low.

Algebra being a fundamental tool in any area amenable to mathematical treatment, these considerations combine to make the algebra of two values of fundamental importance to computer hardware, mathematical logic, and set theory.

Two-valued logic can be extended to multi-valued logic, notably by replacing the Boolean domain {0, 1} with the unit interval [0,1], in which case rather than only taking values 0 or 1, any value between and including 0 and 1 can be assumed. Algebraically, negation (NOT) is replaced with $1 - x$, conjunction (AND) is replaced with multiplication (xy), and disjunction (OR) is defined via De Morgan's law. Interpreting these values as logical truth values yields a multi-valued logic, which forms the basis for fuzzy logic and probabilistic logic. In these interpretations, a value is interpreted as the "degree" of truth – to what extent a proposition is true, or the probability that the proposition is true.

Boolean Operations

The original application for Boolean operations was mathematical logic, where it combines the truth values, true or false, of individual formulas. Natural languages such as English have words for several Boolean operations, in particular conjunction (*and*), disjunction (*or*), negation (*not*), and implication (*implies*). *But not* is synonymous with *and not*. When used to combine situational assertions such as "the block is on the table" and "cats drink milk," which naively are either true or false, the meanings of these logical connectives often have the meaning of their logical counterparts. However with descriptions of behaviour such as "Jim walked through the door", one starts to notice differences such as failure of commutativity, for example the conjunction of "Jim opened the door" with "Jim walked through the door" in that order is not equivalent to their conjunction in the other order, since *and* usually means *and then* in such cases. Questions can be similar: the order "Is the sky blue, and why is the sky blue?" makes more sense than the reverse order. Conjunctive commands about behaviour are like behavioural assertions, as in *get dressed and go to school*. Disjunctive commands such *love me or leave me* or *fish or cut bait* tend to be asymmetric via the implication that one alternative is less preferable. Conjoined nouns such as *tea and milk* generally describe aggregation as with set union while *tea or milk* is a choice. However context can reverse these senses, as in *your choices are coffee and tea* which usually means the same as *your choices are coffee or tea* (alternatives). Double negation as in "I don't not like milk" rarely means literally "I do like milk" but rather conveys some sort of hedging, as though to imply that there is a third possibility. "Not not P" can be loosely interpreted as "surely P", and although P necessarily implies "not not P" the converse is suspect in English, much as with intuitionistic logic. In view of the highly idiosyncratic usage of conjunctions in natural languages, Boolean algebra cannot be considered a reliable framework for interpreting them.

Boolean operations are used in digital logic to combine the bits carried on individual wires, thereby interpreting them over {0,1}. When a vector of n identical binary gates are used to combine two bit vectors each of n bits, the individual bit operations can be understood collectively as a single operation on values from a Boolean algebra with 2^n elements.

Naive set theory interprets Boolean operations as acting on subsets of a given set X. As we saw earlier this behaviour exactly parallels the coordinate-wise combinations of bit vectors, with the union of two sets corresponding to the disjunction of two bit vectors and so on.

The 256-element free Boolean algebra on three generators is deployed in computer displays based on raster graphics, which use bit blit to manipulate whole regions consisting of pixels, relying on Boolean operations to specify how the source region should be combined with the destination, typically with the help of a third region called the mask. Modern video cards offer all $2^{23} = 256$ ternary operations for this purpose, with the choice of operation being a one-byte (8-bit) parameter. The constants SRC = 0xaa or 10101010, DST = 0xcc or 11001100, and MSK = 0xf0 or 11110000 allow Boolean operations such as (SRC^DST)&MSK (meaning XOR the source and destination and then AND the result with the mask) to be written directly as a constant denoting a byte calculated at compile time, 0x60 in the (SRC^DST)&MSK example, 0x66 if just SRC^DST, etc. At run time the video card interprets the byte as the raster operation indicated by the original expression in a uniform way that requires remarkably little hardware and which takes time completely independent of the complexity of the expression.

Solid modelling systems for computer aided design offer a variety of methods for building objects from other objects, combination by Boolean operations being one of them. In this method the space in which objects exist is understood as a set S of voxels (the three-dimensional analogue of pixels in two-dimensional graphics) and shapes are defined as subsets of S, allowing objects to be combined as sets via union, intersection, etc. One obvious use is in building a complex shape from simple shapes simply as the union of the latter. Another use is in sculpting understood as removal of material: any grinding, milling, routing, or drilling operation that can be performed with physical machinery on physical materials can be simulated on the computer with the Boolean operation $x \wedge \neg y$ or $x - y$, which in set theory is set difference, remove the elements of y from those of x. Thus given two shapes one to be machined and the other the material to be removed, the result of machining the former to remove the latter is described simply as their set difference.

Chapter 4

Sequent Calculus

In proof theory and mathematical logic, sequent calculus is a family of formal systems sharing a certain style of inference and certain formal properties. The first sequent calculi, systems LK and LJ, were introduced by Gerhard Gentzen in 1934 as a tool for studying natural deduction in first-order logic (in classical and intuitionistic versions, respectively). Gentzen's so-called "Main Theorem" (*Hauptsatz*) about LK and LJ was the cut-elimination theorem, a result with far-reaching meta-theoretic consequences, including consistency. Gentzen further demonstrated the power and flexibility of this technique a few years later, applying a cut-elimination argument to give a (transfinite) proof of the consistency of Peano arithmetic, in surprising response to Gödel's incompleteness theorems. Since this early work, sequent calculi (also called Gentzen systems) and the general concepts relating to them have been widely applied in the fields of proof theory, mathematical logic, and automated deduction.

One way to classify different styles of deduction systems is to look at the form of *judgements* in the system, *i.e.*, which things may appear as the conclusion of a (sub)proof. The simplest judgement form is used in Hilbert-style deduction systems, where a judgement has the form B where $B$is any formula of first-order-logic (or whatever logic the deduction system applies to, *e.g.*, propositional calculus or a higher-order logic or a modal logic). The theorems are those formulae that appear as the concluding judgement in a valid proof. A Hilbert-style system needs no distinction between formulae and judgements; we make one here solely for comparison with the cases that follow.

The price paid for the simple syntax of a Hilbert-style system is that complete formal proofs tend to get extremely long. Concrete arguments about proofs in such a system almost always appeal to the deduction theorem. This leads to the idea of including the deduction theorem as a formal rule in the system, which happens in natural deduction. In natural deduction, judgements have the shape

$$A_1, A_2, \ldots, A_n \vdash B$$

where the A_i 's $B$and are again formulae and $n \geq 0$. In words, a judgement consists of a list (possibly empty) of formulae on the left-hand side of a turnstile symbol "$\vdash$", with a single formula on the right-hand side. The theorems are those formulae Bsuch that $\vdash B$ (with an empty left-hand side) is the conclusion of a valid proof. (In some presentations of natural deduction, the A_is and the turnstile are not written down explicitly; instead a two-dimensional notation from which they can be inferred is used).

The standard semantics of a judgement in natural deduction is that it asserts that whenever A_1, A_2, etc., are all true, Bwill also be true. The judgements

$$A_1, \ldots, A_n \vdash B \quad \text{and} \quad \vdash (A_1 \wedge \cdots \wedge A_n) \to B$$

are equivalent in the strong sense that a proof of either one may be extended to a proof of the other.

Finally, *sequent calculus* generalizes the form of a natural deduction judgement to

$$A_1, \ldots, A_n \vdash B_1, \ldots, B_k,$$

a syntactic object called a sequent. The formulas on left-hand side of the turnstile are called the *antecedent*, and the formulas on right-hand side are called the *succedent*; together they are called *cedents*. Again, A_i and B_i are formulae, and $n$and $k$are nonnegative integers, that is, the left-hand-side or the right-hand-side (or neither or both) may be empty. As in natural deduction, theorems are those $B$where $\vdash B$ is the conclusion of a valid proof. The empty sequent, having both cedents empty, is defined to be false.

The standard semantics of a sequent is an assertion that whenever *every* is A_i true, *at least one* B_i will also be true. One way to express this is that a comma to the left of the turnstile should be thought of as an "and", and a comma to the right of the turnstile should be thought of as an (inclusive) "or". The sequents

$$A_1,\ldots,A_n \vdash B_1,\ldots,B_k \quad \text{and} \quad \vdash (A_1 \wedge \cdots \wedge A_n) \rightarrow (B_1 \vee \cdots \vee B_k)$$

are equivalent in the strong sense that a proof of either one may be extended to a proof of the other.

At first sight, this extension of the judgement form may appear to be a strange complication — it is not motivated by an obvious shortcoming of natural deduction, and it is initially confusing that the comma seems to mean entirely different things on the two sides of the turnstile. However, in a classical context the semantics of the sequent can also (by propositional tautology) be expressed either as

$$\vdash \neg A_1 \vee \neg A_2 \vee \cdots \vee \neg A_n \vee B_1 \vee B_2 \vee \cdots \vee B_k$$

(at least one of the As is false, or one of the Bs is true) or as

$$\vdash \neg(A_1 \wedge A_2 \wedge \cdots \wedge A_n \wedge \neg B_1 \wedge \neg B_2 \wedge \cdots \wedge \neg B_k)$$

(it cannot be the case that all of the As are true and all of the Bs are false). In these formulations, the only difference between formulae on either side of the turnstile is that one side is negated. Thus, swapping left for right in a sequent corresponds to negating all of the constituent formulae. This means that a symmetry such as De Morgan's laws, which manifests itself as logical negation on the semantic level, translates directly into a left-right symmetry of sequents — and indeed, the inference rules in sequent calculus for dealing with conjunction ($\wedge$) are mirror images of those dealing with disjunction ($\vee$).

Many logicians feel that this symmetric presentation offers a deeper insight in the structure of the logic than other styles of proof system, where the classical duality of negation is not as apparent in the rules.

The System LK

This section introduces the rules of the sequent calculus LK (which is short for "logistischer klassischer Kalkül"), as introduced by Gentzen in 1934. A (formal) proof in this calculus is a sequence of sequents, where each of the sequents is derivable from sequents appearing earlier in the sequence by using one of the rules below.

An Intuitive Explanation

The above rules can be divided into two major groups: *logical* and *structural* ones. Each of the logical rules introduces a new logical formula either on the left or on the right of the turnstile $\vdash$. In contrast, the structural rules operate on the structure of the sequents, ignoring the exact shape of the formulae. The two exceptions to this general scheme are the axiom of identity (I) and the rule of (Cut).

Although stated in a formal way, the above rules allow for a very intuitive reading in terms of classical logic. Consider, for example, the rule ($\wedge L_1$). It says that, whenever one can prove that Δ can be concluded from some sequence of formulae that contain A, then one can also conclude Δ from the (stronger) assumption, that $A \wedge B$ holds. Likewise, the rule ($\neg R$) states that, if Γ and A suffice to conclude Δ, then from Γ alone one can either still conclude Δ or A must be false, i.e. $\neg A$ holds. All the rules can be interpreted in this way.

For an intuition about the quantifier rules, consider the rule ($\forall$ R). Of course concluding that $\forall$ x A holds just from the fact that A[y/x] is true is not in general possible. If, however, the variable y is not mentioned elsewhere (i.e. it can still be chosen freely, without influencing the other formulae), then one may assume, that A[y/x] holds for any value of y. The other rules should then be pretty straightforward.

Instead of viewing the rules as descriptions for legal derivations in predicate logic, one may also consider them as instructions for the construction of a proof for a given statement. In this case the rules can be read bottom-up; for example, ($\wedge$R) says that, to prove that $A \wedge B$ follows from the assumptions Γ and Σ, it suffices to prove that A can be concluded from Γ and B can be concluded from Σ, respectively. Note that, given some antecedent, it is not clear how this is to be split into Γ and Σ. However, there are only finitely many possibilities to be checked since the antecedent by assumption is finite. This also illustrates how proof theory can be viewed as operating on proofs in a combinatorial fashion: given proofs for both A and B, one can construct a proof for $A \wedge B$.

When looking for some proof, most of the rules offer more or less direct recipes of how to do this. The rule of cut is different: It states that, when a formula A can be concluded and this formula may also serve as a premise for concluding other statements, then the formula A can be "cut out" and the respective derivations are joined. When constructing a proof bottom-up, this creates the problem of guessing A (since it does not appear at all below). The cut-elimination theorem is thus crucial to the applications of sequent calculus in automated deduction: it states that all uses of the cut rule can be eliminated from a proof, implying that any provable sequent can be given a *cut-free* proof.

The second rule that is somewhat special is the axiom of identity (I). The intuitive reading of this is obvious: every formula proves itself. Like the cut rule, the axiom of identity is somewhat redundant: the completeness of atomic initial sequents states that the rule can be restricted to atomic formulas without any loss of provability.

Observe that all rules have mirror companions, except the ones for implication. This reflects the fact that the usual language of first-order logic does not include the "is not implied by" connective $\not\leftarrow$ that would be the De Morgan dual of implication. Adding such a connective with its natural rules would make the calculus completely left-right symmetric.

Structural Rules

The structural rules deserve some additional discussion. Weakening (W) allows the addition of arbitrary elements to a sequence. Intuitively, this is allowed in the antecedent because we can always restrict the scope of our proof (if all cars have wheels, then it's safe to say that all black cars have wheels); and in the succedent because we can always allow for alternative conclusions (if all cars have wheels, then it's safe to say that all cars have either wheels or wings).

Contraction (C) and Permutation (P) assure that neither the order (P) nor the multiplicity of occurrences (C) of elements of the sequences matters. Thus, one could instead of sequences also consider sets.

The extra effort of using sequences, however, is justified since part or all of the structural rules may be omitted. Doing so, one obtains the so-called substructural logics.

Properties of the System LK

This system of rules can be shown to be both sound and complete with respect to first-order logic, i.e. a statement Afollows semantically from a set of premises $\Gamma(\Gamma \models A)$ iff the sequent $\Gamma \vdash A$ can be derived by the above rules.

In the sequent calculus, the rule of cut is admissible. This result is also referred to as Gentzen's *Hauptsatz* ("Main Theorem").

Variants

The above rules can be modified in various ways:

Minor Structural Alternatives

There is some freedom of choice regarding the technical details of how sequents and structural rules are formalized. As long as every derivation in LK can be effectively transformed to a derivation using the new rules and vice versa, the modified rules may still be called LK.

First of all, as mentioned above, the sequents can be viewed to consist of sets or multisets. In this case, the rules for permuting and (when using sets) contracting formulae are obsolete.

The rule of weakening will become admissible, when the axiom (I) is changed, such that any sequent of the form $\Gamma, A \vdash A, \Delta$ can be concluded. This means that Aproves in any context. Any weakening that appears in a derivation can then be performed right at the start. This may be a convenient change when constructing proofs bottom-up.

Independent of these one may also change the way in which contexts are split within the rules: In the cases ($\wedge$R), ($\vee$L), and ($\rightarrow$L) the left context is somehow split into Γ and Σ when going upwards. Since contraction allows for the duplication of these, one may assume that the full context is used in both branches of the derivation. By doing this, one assures that no important premises are lost in the wrong branch. Using weakening, the irrelevant parts of the context can be eliminated later.

Substructural Logics

Alternatively, one may restrict or forbid the use of some of the structural rules. This yields a variety of substructural logic systems. They are generally weaker than LK (*i.e.*, they have fewer theorems), and thus not complete with respect to the standard semantics of first-order logic. However, they have other interesting properties that have led to applications in theoretical computer science and artificial intelligence.

Intuitionistic Sequent Calculus: System LJ

Surprisingly, some small changes in the rules of LK suffice to turn it into a proof system for intuitionistic logic. To this end, one has to restrict to sequents with exactly one formula on the right-hand side, and modify the rules to maintain this invariant. For example, (∨L) is reformulated as follows (where C is an arbitrary formula):

$$\frac{\Gamma, A \vdash C \qquad \Sigma, B \vdash C}{\Gamma, \Sigma, A \vee B \vdash C} \ (\vee L)$$

The resulting system is called LJ. It is sound and complete with respect to intuitionistic logic and admits a similar cut-elimination proof.

Boolean Circuit

In computational complexity theory and circuit complexity, a Boolean circuit is a mathematical model for digital logic circuits. A formal language can be decided by a family of Boolean circuits, one circuit for each possible input length. Boolean circuits are also used as a formal model for combinational logic in digital electronics.

Boolean circuits are defined in terms of the logic gates they contain. For example, a circuit might contain binary AND and OR gates and unary NOT gates, or be entirely described by binary NAND gates. Each gate corresponds to some Boolean function that takes a fixed number of bits as input and outputs a single bit.

Boolean circuits provide a model for many digital components used in computer engineering, including multiplexers, adders, and arithmetic logic units.

Formal Definition

In giving a formal definition of Boolean circuits, Vollmer starts by defining a basis set B of Boolean functions, corresponding to the gates allowable in the circuit model. A

Boolean circuit over a basis B, with n inputs and m outputs, is then defined as a finite directed acyclic graph. Each vertex corresponds to either a basis function or one of the inputs, and there are a set of exactly m nodes which are labelled as the outputs. The edges must also have some ordering, to distinguish between different arguments to the same Boolean function.

A common basis for Boolean circuits is the set {AND, OR, NOT}, from which all other Boolean functions can be constructed.

Computational Complexity

Evaluation of a Circuit: The Circuit Value Problem, the problem of computing the output of a given Boolean circuit on a given input string, is a P-complete decision problem. Therefore, this problem is considered to be "inherently sequential" in the sense that there is likely no efficient, highly parallel algorithm that solves the problem.

Complexity Measures

Several important complexity measures can be defined on Boolean circuits, including circuit depth, circuit size, and number of alternations between AND gates and OR gates. For example, the size complexity of a Boolean circuit is the number of gates.

Circuit Complexity

In theoretical computer science, circuit complexity is a branch of computational complexity theory in which Boolean functions are classified according to the size or depth of Boolean circuits that compute them. One speaks of the circuit complexity of a Boolean circuit. A related notion is the circuit complexity of a recursive language that is decided by a family of circuits $C_1, C_2, \ldots$.

A Boolean circuit with $n$input bits is a directed acyclic graph in which every node (usually called *gates* in this context) is either an input node of in-degree 0 labelled by one of the input bits, an AND gate, an OR or a NOT gate. One of these gates is designated as the output gate. Such a circuit naturally computes a function of its inputs. The size of a circuit is the number of gates it contains and its depth is the maximal length of a path from an input gate to the output gate.

There are two major notions of circuit complexity (these are outlined in Sipser (1997)). The circuit-size complexity of a Boolean function f is the minimal size of any circuit computing f. The circuit-depth complexity of a Boolean function f is the minimal depth of any circuit computing f.

These notions generalize when one considers the circuit complexity of a recursive language: A formal language may contain strings with many different bit lengths. Boolean circuits, however, only allow a fixed number of input bits. Thus no single Boolean circuit is capable of deciding such a language. To account for this possibility, one considers families of circuits $C_1, C_2, \ldots$ where each C_n accepts inputs of size n. Each circuit family will naturally generate a recursive language by outputting 1when a string is a member of the family, and 0 otherwise. We say that a family of circuits is size minimal if there is no other family that decides on inputs of any size, n, with a circuit of smaller size than C_n (respectively for depth minimal families).

Hence, the circuit-size complexity of a recursive language $A$is defined as the function $t : \mathbb{N} \rightarrow \mathbb{N}$, that relates a bit length of an input, n, to the circuit-size complexity of the minimal circuit C_n that decides whether inputs of that length are in A. The circuit-depth complexity is defined similarly.

Complexity classes defined in terms of Boolean circuits include AC^0, AC, TC^0 and NC.

Uniformity

Boolean circuits are one of the prime examples of so-called non-uniform models of computation in the sense that inputs of different lengths are processed by different circuits, in contrast with uniform models such as Turing machines where the same computational device is used for all possible input lengths. An individual computational problem is thus associated with a particular *family* of Boolean circuits $C_1, C_2, \ldots$ where each C_n is the circuit handling inputs of n bits. A *uniformity* condition is often imposed on these families, requiring the existence of some resource-bounded Turing machine which, on input n, produces a description of the individual circuit C_n. When this

Turing machine has a running time polynomial in n, the circuit family is said to be P-uniform. The stricter requirement of DLOGTIME-uniformity is of particular interest in the study of shallow-depth circuit-classes such as AC^0 or TC^0.

Polynomial-time Uniform

A family of Boolean circuits $\{C_n : n \in \mathbb{N}\}$ is *polynomial-time uniform* if there exists a deterministic Turing machine *M*, such that

- *M* runs in polynomial time
- For all $n \in \mathbb{N}$, *M* outputs a description of C_n on input 1^n

Logspace Uniform

A family of Boolean circuits $\{C_n : n \in \mathbb{N}\}$ is *logspace uniform* if there exists a deterministic Turing machine *M*, such that

- *M* runs in logarithmic space
- For all $n \in \mathbb{N}$, *M* outputs a description of C_n on input 1^n

History

Circuit complexity goes back to Shannon (1949), who proved that almost all Boolean functions on n variables require circuits of size $\Theta(2^n/n)$. Despite this fact, complexity theorists have not been able to prove superpolynomial circuit lower bounds for specific Boolean functions.

On the other hand, superpolynomial lower bounds have been proved under certain restrictions on the family of circuits used. The first function for which superpolynomial circuit lower bounds were shown was the parity function, which computes the sum of its input bits modulo 2. The fact that parity is not contained in AC^0 was first established independently by Ajtai (1983) and by Furst, Saxe and Sipser (1984). Later improvements by Håstad (1987) in fact establish that any family of constant-depth circuits computing the parity function requires exponential size. Smolensky (1987) proved that this is true even if the circuit is augmented with gates computing the sum of its input bits modulo some odd prime p.

The k-clique problem is to decide whether a given graph on n vertices has a clique of size k. For any particular choice of the constants n and k, the graph can be encoded in binary using $\binom{n}{2}$ bits which indicate for each possible edge whether it is present. Then the k-clique problem is formalized as a function $f_k : \{0,1\}^{\binom{n}{2}} \to \{0,1\}$ such that f_k outputs *1* if and only if the graph encoded by the string contains a clique of size k. This family of functions is monotone and can be computed by a family of circuits, but it has been shown that it cannot be computed by a polynomial-size family of monotone circuits (that is, circuits with AND and OR gates but without negation). The original result of Razborov (1985) was later improved to an exponential-size lower bound by Alon and Boppana (1987). Rossman (2008) shows that constant-depth circuits with AND, OR, and NOT gates require size $\Omega(n^{k/4})$ to solve the k-clique problem even in the average case. Moreover, there is a circuit of size $n^{k/4+O(1)}$ which computes f_k.

Raz and McKenzie later showed that the monotone NC hierarchy is infinite (1999).

The Integer Division Problem lies in uniform TC^0 (Hesse 2001).

Circuit Lower Bounds

Circuit lower bounds are generally difficult. Known results include

- Parity is not in nonuniform AC^0, proved by Ajtai (1983) and by Furst, Saxe and Sipser.
- Uniform TC^0 is not contained in PP, proved by Allender.
- The classes SP

2, PP and MA/1 (MA with one bit of advice) are not in SIZE(n^k) for any constant k.

- While it is suspected that the nonuniform class ACC^0 does not contain the majority function, it was only in 2010 that Williams proved that $\text{NEXP} \not\subseteq \text{ACC}^0$.

It is open whether NEXPTIME has nonuniform TC^0 circuits.

Proofs of circuit lower bounds are strongly connected to derandomization. A proof that P = BPP would imply that either

$\mathsf{NEXP} \not\subseteq \mathsf{P/poly}$ or that permanent cannot be computed by nonuniform arithmetic circuits (polynomials) of polynomial size and polynomial degree.

Complexity Classes

Many circuit complexity classes are defined in terms of class hierarchies. For each nonnegative integer *i*, there is a class NC^i, consisting of polynomial-size circuits of depth $O(\log^i(n))$, using bounded fan-in AND, OR, and NOT gates. We can talk about the union NC of all of these classes. By considering unbounded fan-in gates, we construct the classes AC^i and AC. We construct many other circuit complexity classes with the same size and depth restrictions by allowing different sets of gates.

Relation to Time Complexity

Say that a certain language, *A*, belongs to the time-complexity class $\mathrm{TIME}(t(n))$ for some function $t : \mathbb{N} \to \mathbb{N}$. Then *A*has circuit complexity $\mathcal{O}(t^2(n))$

Functions and Relations

A *relation* is any association between elements of one set, called the *domain* or (less formally) the *set of inputs*, and another set, called the *range* or *set of outputs*. Some people mistakenly refer to the range as the *codomain*, but as we will see, that really means the *set of all possible outputs*—even values that the relation does not actually use.

For example, if the *domain* is a set Fruits = {apples, oranges, bananas} and the *codomain* is a set Flavors = {sweetness, tartness, bitterness}, the flavors of these fruits form a relation: we might say that apples are related to (or associated with) both sweetness and tartness, while oranges are related to tartness only and bananas to sweetness only. (We might disagree somewhat, but that is irrelevant to the topic of this book.) Notice that "bitterness", although it is one of the possible Flavors (codomain), is not really used for any of these relationships; so it is not part of the *range* {sweetness, tartness}.

Another way of looking at this is to say that a relation is a *subset of ordered pairs* drawn from the *set of all possible ordered pairs* (of elements of two other sets, which we normally refer to as the *Cartesian product* of those sets). Formally, R is a relation if

$$R \subseteq \{(x, y) \mid x \in X, y \in Y\}$$

for the domain X and codomain Y.

Using the example above, we can write the relation in set notation: {(apples, sweetness), (apples, tartness), (oranges, tartness), (bananas, sweetness)}.

One important kind of relation is the *function*. A *function* is a relation that has exactly one output for every possible input in the domain. (Unlike the codomain, the domain does not necessarily have to include all possible objects of a given type. In fact, we sometimes intentionally use a *restricted domain* in order to satisfy some desirable property.) For example, the relation that we discussed above (flavors of fruits) is not a function, because it has two possible outputs for the input "apples": sweetness and tartness.

The main reason for not allowing multiple outputs with the same input is that it lets us apply the same function to different forms of the same thing without changing their equivalence. That is, if $x = y$, and f is a function with x (or y) in its domain, then $f(x) = f(y)$. For example, $z - 3 = 5$ implies that $z = 8$ because $f(x) = x + 3$ is a function defined for all numbers x.

The converse, that $f(x) = f(y)$ implies $x = y$, is not always true. When it is, f is called a one-to-one or invertible function.

Relations

In the above section dealing with functions and their properties, we noted the important property that all functions must have, namely that if a function does map a value from its domain to its co-domain, it must map this value to only one value in the co-domain.

Writing in set notation, if *a* is some fixed value:

$$|\{f(x) \mid x=a\}| \in \{0, 1\}$$

The literal reading of this statement is: the *cardinality* (number of elements) of the set of all values f(x), such that x=a for some fixed value a, is an element of the set {0, 1}. In other words, the number of *outputs* that a function f may have at any fixed *input* a is either zero (in which case it is *undefined* at that input) or one (in which case the output is unique).

However, when we consider the *relation*, we relax this constriction, and so a relation may map one value to more than one other value. In general, a relation is any subset of the Cartesian product of its domain and co-domain.

All functions, then, can be considered as relations also.

Notations

When we have the property that one value is related to another, we call this relation a *binary relation* and we write it as

x R y

where R is the relation.

For arrow diagrams and set notations, remember for relations we do not have the restriction that functions do and we can draw an arrow to represent the mappings, and for a set diagram, we need only write all the ordered pairs that the relation does take: again, by example

f = {(0,0),(1,1),(1,-1),(2,2),(2,-2)}

is a relation and not a function, since both 1 and 2 are mapped to two values, 1 and -1, and 2 and -2 respectively) example let A=2,3,5;B=4,6,9 then A*B=(2,4),(2,6),(2,9),(3,4),(3,6),(3,9),(5,4),(5,6),(5,9) Define a relation R=(2,4),(2,6),(3,6),(3,9) add functions and problems to one another

Some Simple Examples

Let us examine some simple relations.

Say f is defined by

{(0,0),(1,1),(2,2),(3,3),(1,2),(2,3),(3,1),(2,1),(3,2),(1,3)}

This is a relation (not a function) since we can observe that 1 maps to 2 and 3, for instance.

Less-than, "<", is a relation also. Many numbers can be less than some other fixed number, so it cannot be a function.

Properties

When we are looking at relations, we can observe some special properties different relations can have.

Reflexive

A relation is *reflexive* if, we observe that for all values a:

a R a

In other words, all values are related to themselves.

The relation of equality, "=" is reflexive. Observe that for, say, all numbers a (the domain is R):

$a = a$

so "=" is reflexive.

In a reflexive relation, we have arrows for all values in the domain pointing back to themselves:

Note that ≤ is also reflexive (a ≤ a for any a in R). On the other hand, the relation < is not (a < a is false for any a in R).

Symmetric

A relation is symmetric *if, we observe that for all values a and b:*

a R b implies b R a

The relation of equality again is symmetric. If $x=y$, we can also write that $y=x$ also.

In a symmetric relation, for each arrow we have also an opposite arrow, i.e. there is either no arrow between x and y, or an arrow points from x to y and an arrow back from y to x:

Neither ≤ nor < is symmetric (2 ≤ 3 and 2 < 3 but not 3 ≤ 2 nor 3 < 2 is true).

Transitive

A relation is *transitive* if for all values a, b, c:

a R b and b R c implies a R c

The relation *greater-than* ">" is transitive. If $x > y$, and $y > z$, then it is true that $x > z$. This becomes clearer when we write down what is happening into words. x is greater than y and y is greater than z. So x is greater than both y and z.

The relation *is-not-equal* "$\neq$" is not transitive. If $x \neq y$ and $y \neq z$ then we might have $x = z$ or $x \neq z$ (for example $1 \neq 2$ and $2 \neq 3$ and $1 \neq 3$ but $0 \neq 1$ and $1 \neq 0$ and $0 = 0$).

In the arrow diagram, every arrow between two values a and b, and b and c, has an arrow going straight from a to c.

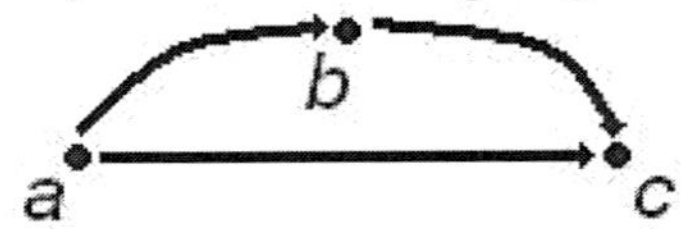

Antisymmetric

A relation is *antisymmetric* if we observe that for all values a and b:

$$a \text{ R } b \text{ and } b \text{ R } a \text{ implies that } a=b$$

Notice that antisymmetric is not the same as "not symmetric."

Take the relation *greater than or equal to*, "$\geq$" If $x \geq y$, and $y \geq$ x, then y must be equal to x. a relation is anti-symmetric if and only if $a \in A$, $(a,a) \in R$

Trichotomy

A relation satisfies trichotomy *if we observe that for all values* a *and* b *it holds true that:* a*R*b or b*R*a

The relation is-greater-or-equal *satisfies since, given 2 real numbers* a *and* b, *it is true that whether* $a \geq b$ *or* $b \geq a$ *(both if* $a = b$*).*

Problem Set

Given the above information, determine which relations are reflexive, transitive, symmetric, or antisymmetric on the following - there may be more than one characteristic. (Answers follow.) x *R* y *if*

1. $x = y$
2. $x < y$
3. $x^2 = y^2$
4. $x \leq y$

Answers

1. Symmetric, Reflexive, Transitive and Antisymmetric
2. Transitive
3. Symmetric, Reflexive, Transitive and Antisymmetric ($x^2 = y^2$ is just a special case of equality, so all properties that apply to $x = y$ also apply to this case)
4. Reflexive, Transitive and Antisymmetric (and satisfying Trichotomy)

Equivalence Relations

We have seen that certain common relations such as "=", and congruence (which we will deal with in the next section) obey some of these rules above. The relations we will deal with are very important in discrete mathematics, and are known as *equivalence relations*. They essentially assert some kind of equality notion, or *equivalence*, hence the name.

Characteristics of Equivalence Relations

For a relation R to be an equivalence relation, *it must have the following properties, viz. R must be:*

- symmetric
- transitive
- reflexive

(A helpful mnemonic, S-T-R)

In the previous problem set you have shown equality, "=", to be reflexive, symmetric, and transitive. So "=" is an equivalence relation.

We denote an equivalence relation, in general, by $x \sim y$.

Example Proof

Say we are asked to prove that "=" is an equivalence relation. We then proceed to prove each property above in turn (Often, the proof of transitivity is the hardest).

- Reflexive: Clearly, it is true that $a = a$ for all values a. Therefore, = is reflexive.
- Symmetric: If $a = b$, it is also true that $b = a$. Therefore, = is symmetric

- Transitive: If $a = b$ and $b = c$, this says that a is the same as b which in turn is the same as c. So a is then the same as c, so $a = c$, and thus = is transitive.

Thus = is an equivalence relation.

Partitions and Equivalence Classes

It is true that when we are dealing with relations, we may find that many values are related to one fixed value.

For example, when we look at the quality of *congruence*, which is that given some number a, a number congruent to a is one that has the same remainder or *modulus* when divided by some number n, as a, which we write

$$a \equiv b \pmod{n}$$

and is the same as writing

$$b = a + kn \text{ for some integer } k.$$

(We will look into congruences in further detail later, but a simple examination or understanding of this idea will be interesting in its application to equivalence relations)

For example, $2 \equiv 0 \pmod 2$, since the remainder on dividing 2 by 2 is in fact 0, as is the remainder on dividing 0 by 2.

We can show that congruence is an equivalence relation (This is left as an exercise, below Hint use the equivalent form of congruence as described above).

However, what is more interesting is that we can group all numbers that are equivalent to each other.

With the relation congruence *modulo* 2 (which is using n=2, as above), or more formally:

$$x \sim y \text{ if and only if } x \equiv y \pmod 2$$

we can group all numbers that are equivalent to each other. Observe:

$$0 \equiv 2 \equiv 4 \equiv \ldots \pmod 2$$

$$1 \equiv 3 \equiv 5 \equiv \ldots \pmod 2$$

This first equation above tells us all the *even* numbers are equivalent to each other under ~, and all the *odd* numbers under ~.

We can write this in set notation. However, we have a special notation. We write:

[0]={0,2,4,...}

[1]={1,3,5,...}

and we call these two sets *equivalence classes*.

All elements in an equivalence class by definition are equivalent to each other, and thus note that we do not need to include [2], since 2 ~ 0.

We call the act of doing this 'grouping' with respect to some equivalence relation *partitioning* (or further and explicitly *partitioning a set S into equivalence classes under a relation* ~). Above, we have partitioned Z into equivalence classes [0] and [1], under the relation of congruence modulo 2.

Problem Set

Given the above, answer the following questions on equivalence relations (Answers follow to even numbered questions)

1. Prove that congruence is an equivalence relation as before.
2. Partition {x | 1 ≤ x ≤ 9} into equivalence classes under the equivalence relation

$$x \sim y \text{ iff } x \equiv y \pmod{6}$$

Answers

2. [0]={6}, [1]={1,7}, [2]={2,8}, [3]={3,9}, [4]={4}, [5]={5}

Partial Orders

We also see that "≥" and "≤" obey some of the rules above. Are these special kinds of relations too, like equivalence relations? Yes, in fact, these relations are specific examples of another special kind of relation which we will describe in this section: the *partial order*.

As the name suggests, this relation gives some kind of ordering to numbers.

Characteristics of Partial Orders

For a relation R to be a partial order, it must have the following three properties, viz R must be:

- reflexive
- antisymmetric
- transitive

(A helpful mnemonic, R-A-T)

We denote a partial order, in general, by $x \preceq y$.

Example Proof

Say we are asked to prove that "≤" is a partial order. We then proceed to prove each property above in turn (Often, the proof of transitivity is the hardest).

Reflexive

Clearly, it is true that $a \leq a$ for all values a. So ≤ is reflexive.

Antisymmetric

If $a \leq b$, and $b \leq a$, then a *must* be equal to b. So ≤ is antisymmetric

Transitive

If $a \leq b$ and $b \leq c$, this says that a is less than b and c. So a is less than c, so $a \leq c$, and thus ≤ is transitive.

Thus ≤ is a partial order.

Problem Set

Given the above on partial orders, answer the following questions

1. Prove that divisibility, |, is a partial order (a | b means that a is a factor of b, i.e., on dividing b by a, no remainder results).
2. Prove the following set is a partial order: $(a, b) \preceq (c, d)$ implies $ab \leq cd$ for a,b,c,d integers ranging from 0 to 5. Answers
3. Simple proof; Formalization of the proof is an optional exercise.

Reflexivity: (a, b) (a, b) since $ab=ab$.

Antisymmetric: (a, b) (c, d) and (c, d) (a, b) since $ab \leq cd$ and $cd \leq ab$ imply $ab=cd$.

Transitive: (a, b) (c, d) and (c, d) (e, f) implies (a, b) (e, f) since $ab \leq cd \leq ef$ and thus $ab \leq ef$

Posets

A partial order imparts some kind of "ordering" amongst elements of a set. For example, we only know that $2 \geq 1$ because of the partial ordering $\geq$.

We call a set A, ordered under a general partial ordering $\preceq$, a *partially ordered set*, or simply just *poset*, and write it (A,).

Terminology

There is some specific terminology that will help us understand and visualize the partial orders.

When we have a partial order $\preceq$, such that a b, we write to say that a but $a \neq b$. We say in this instance that a *precedes* b, or a is a predecessor of b.

If (A,) is a poset, we say that a is an immediate predecessor of b (or a immediately precedes b) if there is no x in A such that a x b.

If we have the same poset, and we also have a and b in A, then we say a and b are *comparable* if a b or b a. Otherwise they are *incomparable*.

Hasse Diagrams

Hasse diagrams *are special diagrams that enable us to visualize the structure of a partial ordering. They use some of the concepts in the previous section to draw the diagram.*

A Hasse diagram of the poset $(A, \preceq)$ is constructed by

- placing elements of A as points
- if a and $b \in$ A, and a is an immediate predecessor of b, we draw a line from a to b
- if $a \prec b$, put the point for a lower than the point for b
- not drawing loops from a to a (this is assumed in a partial order because of reflexivity)

Operations on Relations

There are some useful operations one can perform on relations, which allow to express some of the above mentioned properties more briefly.

Inversion

Let R be a relation, then its inversion, R^{-1} is defined by

R^{-1} := {(a,b) | (b,a) in R}.

Concatenation

Let R be a relation between the sets A and B, S be a relation between B and C. We can concatenate these relations by defining

R • S := {(a,c) | (a,b) in R and (b,c) in S for some b out of B}

Diagonal of a Set

Let A be a set, then we define the diagonal (D) of A by

D(A) := {(a,a) | a in A}

Shorter Notations

Using above definitions, one can say (lets assume R is a relation between A and B):

R is transitive *if and only if R • R is a subset of R.*

R is reflexive *if and only if D(A) is a subset of R.*

R is symmetric *if R^{-1} is a subset of R.*

R is antisymmetric *if and only if the intersection of R and R^{-1} is D(A).*

R is asymmetric *if and only if the intersection of D(A) and R is empty.*

R is a function *if and only if R^{-1} • R is a subset of D(B).*

In this case it is a function A → B. Let's assume R meets the condition of being a function, then

R is injective *if R • R^{-1} if a subset of D(A).*

R is surjective *if {b | (a,b) in R} = B.*

Functions

A function is a relationship between two sets of numbers. We may think of this as a *mapping*; a function *maps* a number in one set to a number in another set. Notice that a function maps values to one and only one value. Two values in one set could map to one value, but one value must never map to two values: that would be a relation, *not* a function.

For example, if we write (define) a function as:

$$f(x) = x^2$$

then we say:

'f of x equals x squared'

and we have

$$f(-1) = 1$$
$$f(1) = 1$$
$$f(7) = 49$$
$$f(1/2) = 1/4$$
$$f(4) = 16$$

and so on.

This function f maps numbers to their squares.

Range, image, codomain

If D is a set, we can say

$$f(D) = \{f(x) \mid x \in D\}$$

which forms a new set, called the *range* of f. D is called the *domain* of f, and represents all values that f takes.

In general, the range of f is usually a subset of a larger set. This set is known as the *codomain* of a function. For example, with the function f(x)=cos x, the range of f is [-1,1], but the codomain is the set of real numbers.

Notations

When we have a function f, with domain D and range R, we write:

$$f : D \to R$$

If we say that, for instance, x is mapped to x^2, we also can add

$$f : D \to R;\ x \to x^2$$

Notice that we can have a function that maps a point (x,y) to a real number, or some other function of two variables — we have a set of ordered pairs as the domain. Recall from set theory that this is defined by the *Cartesian product* - if we wish to represent a set of all real-valued ordered pairs we can take the Cartesian product of the real numbers with itself to obtain

$$\mathbb{R} \times \mathbb{R} = \mathbb{R}^2 = \{(x, y) \mid x \text{ and } y \in \mathbb{R}\}.$$

When we have a set of n-tuples as part of the domain, we say that the function is n-ary (for numbers n=1,2 we say unary, and binary respectively).

Other Function Notation

Functions can be written as above, but we can also write them in two other ways. One way is to use an arrow diagram to represent the mappings between each element. We write the elements from the domain on one side, and the elements from the range on the other, and we draw arrows to show that an element from the domain is mapped to the range.

For example, for the function f(x)=x^3, the arrow diagram for the domain {1,2,3} would be:

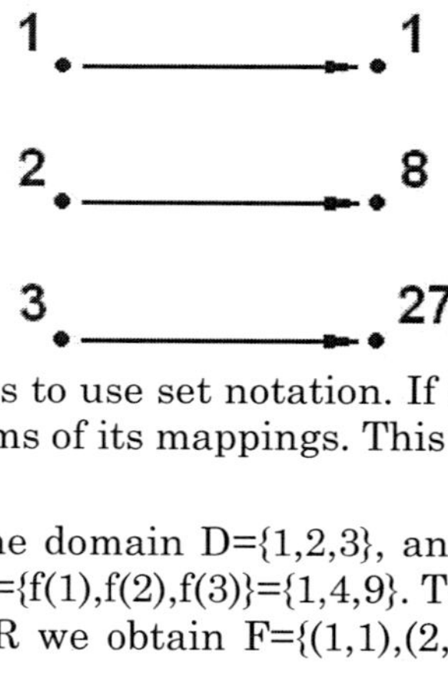

Another way is to use set notation. If f(x)=y, we can write the function in terms of its mappings. This idea is best to show in an example.

Let us take the domain D={1,2,3}, and f(x)=x^2. Then, the range of f will be R={f(1),f(2),f(3)}={1,4,9}. Taking the Cartesian product of D and R we obtain F={(1,1),(2,4),(3,9)}.

So using set notation, a function can be expressed as the Cartesian product of its domain and range.

f(x)

This function is called f, and it takes a *variable* x. We substitute some value for x to get the second value, which is what the function maps x to.

Combinatorics

Combinatorics is a branch of mathematics concerning the study of finite or countable discrete structures. Aspects of combinatorics include counting the structures of a given kind and size (enumerative combinatorics), deciding when certain criteria can be met, and constructing and analyzing objects meeting the criteria (as in combinatorial designs and matroid theory), finding “largest”, “smallest”, or “optimal” objects (extremal combinatorics and combinatorial optimization), and studying combinatorial structures arising in an algebraic context, or applying algebraic techniques to combinatorial problems (algebraic combinatorics).

Combinatorial problems arise in many areas of pure mathematics, notably in algebra, probability theory, topology, and geometry, and combinatorics also has many applications in optimization, computer science, ergodic theory and statistical physics. Many combinatorial questions have historically been considered in isolation, giving an *ad hoc* solution to a problem arising in some mathematical context.

In the later twentieth century, however, powerful and general theoretical methods were developed, making combinatorics into an independent branch of mathematics in its own right. One of the oldest and most accessible parts of combinatorics is graph theory, which also has numerous natural connections to other areas. Combinatorics is used frequently in computer science to obtain formulas and estimates in the analysis of algorithms.

A mathematician who studies combinatorics is called a *combinatorialist.*

History of Combinatorics

The history of combinatorics is an area of study within the history of mathematics, dedicated to the history of combinatorics and its variations, from antiquity to modern times.

Earliest Records

The earliest known connection to combinatorics comes from the Rhind papyrus, problem 79, for the implementation of a geometric series.

A Jain text, the Bhagabati Sutra, had the first mention of a combinatorics problem; it asked how many ways one could take six tastes one, two, or three tastes at a time. The Bhagabati Sutra was written around 300 BC, and was the first book to mention the choose function. The next ideas of Combinatorics came from Pingala, who was interested in prosody. Specifically, he wanted to know how many ways a six-syllable metre could be made from short and long notes. He wrote this problem in the Chanda sutra (also Chandahsutra) in the second century BC. In addition, he also found the number of metres that had n long notes and k short notes, which is equivalent to finding the binomial coefficients.

The ideas of the Bhagabati were generalised by the Indian mathematician Mahavira in 850 AD, and Pingala's work on prosody was expanded by Bhaskara and Hemacandra in 1100 AD. Bhaskara was the first known person to find the generalised choice function, although Brahmagupta may have known earlier. Hemacandra asked how many metres existed of a certain length if a long note was considered to be twice as long as a short note, which is equivalent to finding the Fibonacci numbers.

The ancient Chinese book of divination, I Ching is about what different hexagrams mean, and to do this one needs to know how many possible hexagrams there were. Since each hexagram is a permutation with repetitions of six lines, where each line can be one of two states, solid or dashed, combinatorics yields the result that there are $2^6 = 64$ hexagrams. A monk also may have counted the number of configurations to a game similar to Go around 700 AD. Although China had relatively few advancements in enumerative combinatorics, they solved a combinatorial design problem, the magic square, around 100 AD.

In Greece, Plutarch wrote that the Xenocrates discovered the number of different syllables possible in the Greek language. This, however, is unlikely because this is one of the few mentions of Combinatorics in Greece. The number they found, 1.002×10^{12} also seems too round to be more than a guess.

Magic squares remained an interest of China, and they began to generalise their original 3×3 square between 900 and 1300 AD. China corresponded with the Middle East about this problem in the 13th century. The Middle East also learned about binomial coefficients from Indian work, and found the connection to polynomial expansion.

Abû Bakr ibn Mu%ammad ibn al $usayn Al-Karaji (c.953-1029) wrote on the binomial theorem and Pascal's triangle. In a now lost work known only from subsequent quotation by al-Samaw'al Al-Karaji introduced the idea of argument by mathematical induction.

The philosopher and astronomer Rabbi Abraham ibn Ezra (c. 1140) counted the permutations with repetitions in vocalization of Divine Name. He also established the symmetry of binomial coefficients, while a closed formula was obtained later by the talmudist and mathematician Levi ben Gerson (better known as Gersonides), in 1321. The arithmetical triangle— a graphical diagram showing relationships among the binomial coefficients— was presented by mathematicians in treatises dating as far back as the 10th century, and would eventually become known as Pascal's triangle. Later, in Medieval England, campanology provided examples of what is now known as Hamiltonian cycles in certain Cayley graphs on permutations.

Combinatorics in the West

Combinatorics came to Europe in the 13th century through two mathematicians, Leonardo Fibonacci and Jordanus de Nemore. Fibonacci's Liber Abaci introduced many of the Arabian and Indian ideas to Europe, including that of the Fibonacci numbers. Jordanus was the first person to arrange the binomial coefficients in a triangle, as he did in proposition 70 of *De Arithmetica.* This was also done in the Middle East in 1265, and China around 1300. Today, this triangle is known as Pascal's triangle.

Pascal's contribution to the triangle that bears his name comes from his work on formal proofs about it, in addition to his connection between it and probability. Together with Leibniz and his ideas about partitions in the 17th century, they are considered the founders of modern combinatorics.

Both Pascal and Leibniz understood that algebra and combinatorics corresponded (aka, binomial expansion was equivalent to the choice function). This was expanded by De Moivre, who found the expansion of a multinomial. De Moivre also found the formula for derangements using the principle of inclusion-exclusion, a method different from Nikolaus Bernoulli, who had found them previously. He managed to approximate the binomial coefficients and factorial. Finally, he found a closed form for the Fibonacci numbers by inventing generating functions.

In the 18th century, Euler worked on problems of combinatorics. In addition to working on several problems of probability which link to combinatorics, he worked on the knights tour, Graeco-Latin square, Eulerian numbers, and others. He also invented graph theory by solving the Seven Bridges of Königsberg problem, which also led to the formation of topology. Finally, he broke ground with partitions by the use of generating functions.

Enumerative Combinatorics

Enumerative combinatorics is an area of combinatorics that deals with the number of ways that certain patterns can be formed. Two examples of this type of problem are counting combinations and counting permutations. More generally, given an infinite collection of finite sets $\{S_i\}$ indexed by the natural numbers, enumerative combinatorics seeks to describe a *counting function* which counts the number of objects in S_n for each n. Although counting the number of elements in a set is a rather broad mathematical problem, many of the problems that arise in applications have a relatively simple combinatorial description. The twelvefold way provides a unified framework for counting permutations, combinations and partitions.

The simplest such functions are *closed formulas*, which can be expressed as a composition of elementary functions such as factorials, powers, and so on. For instance, the number of different possible orderings of a deck of n cards is $f(n) = n!$. Often, no closed form is initially available. In these cases, we frequently first derive a recurrence relation, then solve the recurrence to arrive at the desired closed form.

Finally, $f(n)$ may be expressed by a formal power series, called its *generating function*, which is most commonly either the ordinary generating function

$$\sum_{n\geq 1} f(n)x^n$$

or the exponential generating function

$$\sum_{n\geq 1} f(n)\frac{x^n}{n!}.$$

Often, a complicated closed formula yields little insight into the behaviour of the counting function as the number of counted objects grows. In these cases, a simple asymptotic approximation may be preferable. A function $g(n)$ is an asymptotic approximation to $f(n)$ if $f(n)/g(n) \to 1$ as $n \to \infty$. In this case, we write $f(n) \sim g(n)$.

Once determined, the generating function yields the information given by the previous approaches. In addition, the various natural operations on generating functions such as addition, multiplication, differentiation, etc., have a combinatorial significance; this allows one to extend results from one combinatorial problem in order to solve others.

Generating Functions

Generating functions are used to describe families of combinatorial objects. Let $\mathcal{F}$ denote the family of objects and let $F(x)$ be its generating function. Then:

$$F(x) = \sum_{n=0}^{\infty} f_n x^n$$

Where f_n denotes the number of combinatorial objects of size n. The number of combinatorial objects of size n is therefore given by the coefficient of x^n.

Some common operation on families of combinatorial objects and its effect on the generating function will now be developed. The exponential generating function is also sometimes used. In this case it would have the form:

$$F(x) = \sum_{n=0}^{\infty} \frac{f_n}{n!} x^n$$

Union

Given two combinatorial families, $\mathcal{F}$ and $\mathcal{G}$ with generating functions $F(x)$ and $G(x)$ respectively, the union of the two families ($\mathcal{F} \cup \mathcal{G}$) has generating function $F(x) + G(x)$.

Pairs

For two combinatorial families as above the Cartesian product (pair) of the two families ($\mathcal{F} \times \mathcal{G}$) has generating function $F(x)G(x)$.

Sequences

A sequence generalizes the idea of the pair as defined above. Sequences are arbitrary Cartesian products of a combinatorial object with itself. Formally:

$$\text{Seq}(\mathcal{F}) = \epsilon \cup \mathcal{F} \cup \mathcal{F} \times \mathcal{F} \cup \mathcal{F} \times \mathcal{F} \times \mathcal{F} \cup \ldots$$

To put the above in words: An empty sequence or a sequence of one element or a sequence of two elements or a sequence of three elements, etc. The generating function would be:

$$1 + F(x) + [F(x)]^2 + [F(x)]^3 + \ldots = \frac{1}{1 - F(x)}$$

Combinatorial Structures

The above operations can now be used to enumerate common combinatorial objects including trees (binary and plane), Dyck paths and cycles. A combinatorial structure is composed of atoms. For example, with trees the atoms would be the nodes. The atoms which compose the object can either be labelled or unlabelled.

Unlabelled atoms are indistinguishable from each other, while labelled atoms are distinct. Therefore, for a combinatorial object consisting of labelled atoms a new object can be formed by simply swapping two or more atoms.

Binary and Plane Trees

Binary and plane trees are examples of an unlabelled combinatorial structure. Trees consist of nodes linked by edges in such a way that there are no cycles. There is generally a node called the root, which has no parent node. In Plane trees each node can have an arbitrary number of children. In binary trees, a special case of plane trees, each node can have either two or no children. Let $\mathcal{P}$ denote the family of all plane trees. Then this family can be recursively defined as follows:

$$\mathcal{P} = \{\bullet\} \times \text{Seq}(\mathcal{P})$$

In this case $\{\bullet\}$ represents the family of objects consisting of one node. This has generating function x. Let $P(x)$ denote the generating function $\mathcal{P}$ Putting the above description in words: A plane tree consists of a node to which is attached an arbitrary number of subtrees, each of which is also a plane tree. Using the operation on families of combinatorial structures developed earlier this translates to a recursive generating function:

$$P(x) = x\frac{1}{1-P(x)}$$

After solving for $P(x)$:

$$P(x) = \frac{1-\sqrt{1-4x}}{2}$$

An explicit formula for the number of plane trees of size n can now be determined by extracting the coefficient of x^n.

$$p_n = [x^n]P(x) = [x^n]\frac{1-\sqrt{1-4x}}{2}$$

$$= [x^n]\frac{1}{2} - [x^n]\frac{1}{2}\sqrt{1-4x}$$

$$= -\frac{1}{2}[x^n]\sum_{k=0}^{\infty}\binom{\frac{1}{2}}{k}(-4x)^k$$

$$= -\frac{1}{2}\binom{\frac{1}{2}}{n}(-4)^n$$

$$= \frac{1}{n}\binom{2n-2}{n-1}$$

Note: The notation $[x^n]$ $f(x)$ refers to the coefficient of x^n in $f(x)$. The series expansion of the square root is based on Newton's generalization of the Binomial theorem. To get from the fourth to fifth line manipulations using the generalized binomial coefficient is needed.

The expression on the last line is equal to the $(n-1)^{\text{th}}$ Catalan number. Therefore $p_n = c_{n-1}$.

Analytic Combinatorics

In mathematics, analytic combinatorics is one of the many techniques of counting combinatorial objects. It uses the internal structure of the objects to derive formulas for their generating functions. This particular theory is due to Philippe Flajolet and Robert Sedgewick.

Classes of combinatorial structures

Consider the problem of distributing objects given by a generating function into a set of n slots, where a permutation group G of degree n acts on the slots to create an equivalence relation of filled slot configurations, and asking about the generating function of the configurations by weight of the configurations with respect to this equivalence relation, where the weight of a configuration is the sum of the weights of the objects in the slots. We will first explain how to solve this problem in the labelled and the unlabelled case and use the solution to motivate the creation of classes of combinatorial structures.

The Pólya enumeration theorem solves this problem in the unlabelled case. Let $f(z)$ be the ordinary generating function (OGF) of the objects, then the OGF of the configurations is given by the substituted cycle index

$$Z(G)(f(z), f(z^2), \ldots, f(z^n)).$$

In the labelled case we use an exponential generating function (EGF) $g(z)$ of the objects and apply the Labelled enumeration theorem, which says that the EGF of the configurations is given by

$$\frac{g(z)^n}{|G|}.$$

We are able to enumerate filled slot configurations using either PET in the unlabelled case or the labelled enumeration theorem in the labelled case. We now ask about the generating function of configurations obtained when there is more than one set of slots, with a permutation group acting on each. Clearly the orbits do not intersect and we may add the respective generating functions.

Suppose, for example, that we want to enumerate unlabelled sequences of length two or three of some objects contained in a set X. There are two sets of slots, the first one containing two slots, and the second one, three slots. The group acting on the first set is E_2, and on the second slot, E_3. We represent this by the following formal power series in X:

$$X^2 / E_2 + X^3 / E_3$$

where the term X^n / G is used to denote the set of orbits under G and $X^n = X \times \ldots \times X$, which denotes in the obvious way the process of distributing the objects from X with repetition into the n slots. Similarly, consider the labelled problem of creating cycles of arbitrary length from a set of labelled objects X. This yields the following series of actions of cyclic groups:

$$X / C_1 + X^2 / C_2 + X^3 / C_3 + X^4 / C_4 + \cdots.$$

Clearly we can assign meaning to any such power series of quotients (orbits) with respect to permutation groups, where we restrict the groups of degree n to the conjugacy classes $\mathrm{Cl}(S_n)$ of the symmetric group S_n, which form a unique factorization domain. (The orbits with respect to two groups from the same conjugacy class are isomorphic.) This motivates the following definition.

A class $\mathcal{C} \in \mathbb{N}[\mathfrak{A}]$ of combinatorial structures is a formal series

$$\mathcal{C} = \sum_{n \geq 1} \sum_{G \in \mathrm{Cl}(S_n)} c_G (X^n / G)$$

where $\mathfrak{A}$ (the "A" is for "atoms") is the set of primes of the UFD $\{\mathrm{Cl}(S_n)\}_{n \geq 1}$ and $c_G \in \mathbb{N}$. In the following we will simplify our notation a bit and write e.g.

$$E_2 + E_3 \text{ and } C_1 + C_2 + C_3 + \cdots.$$

for the classes mentioned above.

The Flajolet–Sedgewick Fundamental Theorem

A theorem in the Flajolet–Sedgewick theory of symbolic combinatorics treats the enumeration problem of labelled and unlabelled combinatorial classes by means of the creation of symbolic operators that make it possible to translate equations involving combinatorial structures directly (and automatically) into equations in the generating functions of these structures.

Let $\mathcal{C} \in \mathbb{N}[\mathfrak{A}]$ be a class of combinatorial structures. The OGF $F(z)$ of $\mathcal{C}(X)$ where X has OGF $f(z)$ and the EGF $G(z)$ of $\mathcal{C}(X)$ where X is labelled with EGF $g(z)$ are given by

$$F(z) = \sum_{n\geq 1} \sum_{G\in \mathrm{Cl}(S_n)} c_G Z(G)(f(z), f(z^2), \ldots, f(z^n))$$

and

$$G(z) = \sum_{n\geq 1} \left(\sum_{G\in \mathrm{Cl}(S_n)} \frac{c_G}{|G|} \right) g(z)^n .$$

In the labelled case we have the additional requirement that X not contain elements of size zero. It will sometimes prove convenient to add one to $G(z)$ to indicate the presence of one copy of the empty set.

It is possible to assign meaning to both $\mathcal{C} \in \mathbb{Z}[\mathfrak{A}]$ (the most common example is the case of unlabelled sets) and $\mathcal{C} \in \mathbb{Q}[\mathfrak{A}]$. To prove the theorem simply apply PET (Pólya enumeration theorem) and the labelled enumeration theorem.

The power of this theorem lies in the fact that it makes it possible to construct operators on generating functions that represent combinatorial classes.

A structural equation between combinatorial classes thus translates directly into an equation in the corresponding generating functions.

Moreover in the labelled case it is evident from the formula that we may replace $g(z)$ by the atom z and compute the resulting operator, which may then be applied to EGFs. We now proceed to construct the most important operators. The reader may wish to compare with the data on the cycle index page.

The sequence Operator $\mathfrak{S}$

This operator corresponds to the class

$$1 + E_1 + E_2 + E_3 + \cdots$$

and represents sequences, i.e. the slots are not being permuted and there is exactly one empty sequence. We have

$$F(z) = 1 + \sum_{n\geq 1} Z(E_n)(f(z), f(z^2), \ldots, f(z^n)) = 1 + \sum_{n\geq 1} f(z)^n = \frac{1}{1-f(z)}$$

and

$$G(z) = 1 + \sum_{n\geq 1}\left(\frac{1}{|E_n|}\right) g(z)^n = \frac{1}{1-g(z)}.$$

The Cycle Operator $\mathfrak{C}$

This operator corresponds to the class

$$C_1 + C_2 + C_3 + \cdots$$

i.e., cycles containing at least one object. We have

$$F(z) = \sum_{n\geq 1} Z(C_n)(f(z), f(z^2), \ldots, f(z^n)) = \sum_{n\geq 1}\frac{1}{n}\sum_{d|n}\varphi(d) f(z^d)^{n/d}$$

or

$$F(z) = \sum_{k\geq 1}\varphi(k)\sum_{m\geq 1}\frac{1}{km} f(z^k)^m = \sum_{k\geq 1}\frac{\varphi(k)}{k}\log\frac{1}{1-f(z^k)}$$

and

$$G(z) = \sum_{n\geq 1}\left(\frac{1}{|C_n|}\right) g(z)^n = \log\frac{1}{1-g(z)}.$$

This operator, together with the set operator $\mathfrak{P}$, and their restrictions to specific degrees are used to compute random permutation statistics.

There are two useful restrictions of this operator, namely to even and odd cycles.

The labelled even cycle operator $\mathfrak{C}_{\text{even}}$ is

$$C_2 + C_4 + C_6 + \cdots$$

which yields

$$G(z) = \sum_{n\geq 1}\left(\frac{1}{|C_{2n}|}\right) g(z)^{2n} = \frac{1}{2}\log\frac{1}{1-g(z)^2}.$$

This implies that the labelled odd cycle operator $\mathfrak{C}_{\text{odd}}$

$$C_1 + C_3 + C_5 + \cdots$$

is given by

$$G(z) = \log\frac{1}{1-g(z)} - \frac{1}{2}\log\frac{1}{1-g(z)^2} = \frac{1}{2}\log\frac{1+g(z)}{1-g(z)}.$$

The Multiset/Set Operator $\mathfrak{M}$ / $\mathfrak{P}$

The series is

$$1 + S_1 + S_2 + S_3 + \cdots$$

i.e., the symmetric group is applied to the slots. This creates multisets in the unlabelled case and sets in the labelled case (there are no multisets in the labelled case because the labels distinguish multiple instances of the same object from the set being put into different slots).

We include the empty set in both the labelled and the unlabelled case.

The unlabelled case is done using the function

$$M(f(z), y) = \sum_{n\geq 0} y^n Z(S_n)(f(z), f(z^2), \ldots, f(z^n))$$

so that

$$\mathfrak{M}(f(z)) = M(f(z), 1).$$

Evaluating $M(f(z),1)$ we obtain

$$F(z) = \exp\left(\sum_{l\geq 1}\frac{f(z^l)}{l}\right).$$

For the labelled case we have

$$G(z) = 1 + \sum_{n\geq 1}\left(\frac{1}{|S_n|}\right)g(z)^n = \sum_{n\geq 0}\frac{g(z)^n}{n!} = \exp g(z).$$

In the labelled case we denote the operator by $\mathfrak{P}$, and in the unlabelled case, by $\mathfrak{M}$.

Procedure

Typically, one starts with the *neutral class* $\mathcal{E}$, containing a single object of size 0 (the *neutral object*, often denoted by ϵ), and one or more *atomic classes* $\mathcal{Z}$, each containing a single object of size 1.

Next, set-theoretic relations involving various simple operations, such as disjoint unions, products, sets, sequences, and multisets define more complex classes in terms of the already defined classes. These relations may be recursive. The elegance of symbolic combinatorics lies in that the set theoretic, or *symbolic*, relations translate directly into *algebraic* relations involving the generating functions.

In this article, we will follow the convention of using script uppercase letters to denote combinatorial classes and the corresponding plain letters for the generating functions (so the class $\mathcal{A}$ has generating function $A(z)$).

There are two types of generating functions commonly used in symbolic combinatorics—ordinary generating functions, used for combinatorial classes of unlabelled objects, and exponential generating functions, used for classes of labelled objects.

It is trivial to show that the generating functions (either ordinary or exponential) for $\mathcal{E}$ and $\mathcal{Z}$ are $E(z)=1$ and $Z(z)=z$, respectively.

The disjoint union is also simple — for disjoint sets $\mathcal{B}$ and $\mathcal{C}$, $\mathcal{A}=\mathcal{B}\cup\mathcal{C}$ implies $A(z)=B(z)+C(z)$.

The relations corresponding to other operations depend on whether we are talking about labelled or unlabelled structures (and ordinary or exponential generating functions).

Combinatorial Sum

The restriction of unions to disjoint unions is an important one; however, in the formal specification of symbolic combinatorics, it is too much trouble to keep track of which sets are disjoint.

Instead, we make use of a construction that guarantees there is no intersection (*be careful, however; this affects the semantics of the operation as well*).

In defining the *combinatorial sum* of two sets $\mathcal{A}$ and $\mathcal{B}$, we mark members of each set with a distinct marker, for example $\circ$ for members of $\mathcal{A}$ and $\bullet$ for members $\mathcal{B}$ of .

The combinatorial sum is then:

$$\mathcal{A}+\mathcal{B}=(\mathcal{A}\times\{\circ\})\cup(\mathcal{B}\times\{\bullet\})$$

This is the operation that formally corresponds to addition.

Unlabelled Structures

With unlabelled structures, an ordinary generating function (OGF) is used.

The OGF of a sequence A_n is defined as

$$A(x) = \sum_{n=0}^{\infty} A_n x^n$$

Product

The product of two combinatorial classes $\mathcal{A}$ and $\mathcal{B}$ is specified by defining the size of an ordered pair as the sum of the sizes of the elements in the pair.

Thus we have for $a \in \mathcal{A}$ and $b \in \mathcal{B}$, $|(a,b)| = |a| + |b|$. This should be a fairly intuitive definition. We now note that the number of elements in $\mathcal{A} \times \mathcal{B}$ of size n is

$$\sum_{k=0}^{n} A_k B_{n-k}.$$

Using the definition of the OGF and some elementary algebra, we can show that

$$\mathcal{A} = \mathcal{B} \times \mathcal{C} \text{ implies } A(z) = B(z) \cdot C(z).$$

Sequence

The *sequence construction*, denoted by $\mathcal{A} = \mathfrak{G}\{\mathcal{B}\}$ is defined as

$$\mathfrak{G}\{\mathcal{B}\} = \mathcal{E} + \mathcal{B} + (\mathcal{B} \times \mathcal{B}) + (\mathcal{B} \times \mathcal{B} \times \mathcal{B}) + \cdots.$$

In other words, a sequence is the neutral element, or an element of $\mathcal{B}$, or an ordered pair, ordered triple, etc.

This leads to the relation

$$A(z) = 1 + B(z) + B(z)^2 + B(z)^3 + \cdots = \frac{1}{1 - B(z)}.$$

Set

The *set* (or *powerset*) *construction*, denoted by $\mathcal{A} = \mathfrak{P}\{\mathcal{B}\}$ is defined as

$$\mathfrak{P}\{\mathcal{B}\} = \prod_{\beta \in \mathcal{B}} (\mathcal{E} + \{\beta\}),$$

which leads to the relation

$$
\begin{aligned}
A(z) &= \prod_{\beta \in \mathcal{B}} (1 + z^{|\beta|}) \\
&= \prod_{n=1}^{\infty} (1 + z^{n})^{B_n} \\
&= \exp\left(\ln \prod_{n=1}^{\infty} (1 + z^{n})^{B_n} \right) \\
&= \exp\left(\sum_{n=1}^{\infty} B_n \ln(1 + z^{n}) \right) \\
&= \exp\left(\sum_{n=1}^{\infty} B_n \cdot \sum_{k=1}^{\infty} \frac{(-1)^{k-1} z^{nk}}{k} \right) \\
&= \exp\left(\sum_{k=1}^{\infty} \frac{(-1)^{k-1}}{k} \cdot \sum_{n=1}^{\infty} B_n z^{nk} \right) \\
&= \exp\left(\sum_{k=1}^{\infty} \frac{(-1)^{k-1} B(z^{k})}{k} \right),
\end{aligned}
$$

where the expansion

$$\ln(1+u) = \sum_{k=1}^{\infty} \frac{(-1)^{k-1} u^{k}}{k}$$

was used to go from line 4 to line 5.

Multiset

The *multiset construction*, denoted $\mathcal{A} = \mathfrak{M}\{\mathcal{B}\}$ is a generalization of the set construction.

In the set construction, each element can occur zero or one times.

In a multiset, each element can appear an arbitrary number of times. Therefore,

$$\mathfrak{M}\{\mathcal{B}\} = \prod_{\beta \in \mathcal{B}} \mathfrak{G}\{\beta\}.$$

This leads to the relation

$$\begin{aligned} A(z) &= \prod_{\beta \in \mathcal{B}} (1 - z^{|\beta|})^{-1} \\ &= \prod_{n=1}^{\infty} (1 - z^n)^{-B_n} \\ &= \exp\left(\ln \prod_{n=1}^{\infty} (1 - z^n)^{-B_n} \right) \\ &= \exp\left(\sum_{n=1}^{\infty} -B_n \ln(1 - z^n) \right) \\ &= \exp\left(\sum_{k=1}^{\infty} \frac{B(z^k)}{k} \right), \end{aligned}$$

where, similar to the above set construction, we expand $\ln(1 - z^n)$, swap the sums, and substitute for the OGF of $\mathcal{B}$.

Other Elementary Constructions

Other important elementary constructions are:

- the *cycle construction* ($\mathfrak{C}\{\mathcal{B}\}$), like sequences except that cyclic rotations are not considered distinct
- *pointing* ($\Theta\mathcal{B}$), in which each member of $\mathcal{B}$ is augmented by a neutral (zero size) pointer to one of its atoms
- *substitution* ($\mathcal{B} \circ \mathcal{C}$), in which each atom in a member of $\mathcal{B}$ is replaced by a member of $\mathcal{C}$.

The derivations for these constructions are too complicated to show here. Here are the results:

Construction	*Generating function*
$\mathcal{A} = \mathfrak{C}\{\mathcal{B}\}$	$A(z) = \sum_{k=1}^{\infty} \frac{\phi(k)}{k} \ln \frac{1}{1 - B(z^k)}$ (where $\phi(k)$ is the Euler totient function)
$\mathcal{A} = \Theta\mathcal{B}$	$A(z) = z\frac{d}{dz}B(z)$
$\mathcal{A} = \mathcal{B} \circ \mathcal{C}$	$A(z) = B(C(z))$

Examples

Many combinatorial classes can be built using these elementary constructions. For example, the class of plane trees (that is, trees embedded in the plane, so that the order of the subtrees matters) is specified by the recursive relation

$$\mathcal{G} = \mathcal{Z} \times \mathfrak{S}\{\mathcal{G}\}.$$

In other words, a tree is a root node of size 1 and a sequence of subtrees. This gives

$$G(z) = \frac{z}{1 - G(z)}$$

we solve for G(z) by multiplying $1 - G(z)$ to get

$$G(z) - G(z)^2 = z$$

subtracting z and solving for G(z) using the quadratic formula gives

$$G(z) = \frac{1 - \sqrt{1 - 4z}}{2}.$$

Another example (and a classic combinatorics problem) is integer partitions. First, define the class of positive integers $\mathcal{I}$, where the size of each integer is its value:

$$\mathcal{I} = \mathcal{Z} \times \mathfrak{S}\{\mathcal{Z}\}$$

The OGF of $\mathcal{I}$ is then

$$I(z) = \frac{z}{1 - z}.$$

Now, define the set of partitions $\mathcal{P}$ as

$$\mathcal{P} = \mathfrak{M}\{\mathcal{I}\}.$$

The OGF of $\mathcal{P}$ is

$$P(z) = \exp\left(I(z) + \frac{1}{2} I(z^2) + \frac{1}{3} I(z^3) + \cdots \right).$$

Unfortunately, there is no closed form for $P(z)$; however, the OGF can be used to derive a recurrence relation, or, using more advanced methods of analytic combinatorics, calculate the asymptotic behaviour of the counting sequence.

Labelled Structures

An object is *weakly labelled* if each of its atoms has a nonnegative integer label, and each of these labels is distinct. An object is (*strongly* or *well*) *labelled*, if furthermore, these labels comprise the consecutive integers $[1 \ldots n]$. *Note: some combinatorial classes are best specified as labelled structures or unlabelled structures, but some readily admit both specifications.* A good example of labelled structures is the class of labelled graphs.

With labelled structures, an exponential generating function (EGF) is used.

The EGF of a sequence A_n is defined as

$$A(x) = \sum_{n=0}^{\infty} A_n \frac{x^n}{n!}.$$

Product

For labelled structures, we must use a different definition for product than for unlabelled structures. In fact, if we simply used the cartesian product, the resulting structures would not even be well labelled. Instead, we use the so-called *labelled product*, denoted $\mathcal{A} \star \mathcal{B}$.

For a pair $\beta \in \mathcal{B}$ and $\gamma \in \mathcal{C}$, we wish to combine the two structures into a single structure. In order for the result to be well labelled, this requires some relabelling of the atoms in β and γ. We will restrict our attention to relabellings that are consistent with the order of the original labels.

Note that there are still multiple ways to do the relabelling; thus, each pair of members determines not a single member in the product, but a set of new members. The details of this construction are found on the page of the Labelled enumeration theorem.

To aid this development, let us define a function, ρ, that takes as its argument a (possibly weakly) labelled object α and relabels its atoms in an order-consistent way so that $\rho(\alpha)$ is well labelled. We then define the labelled product for two objects α and β as

$$\alpha \star \beta = \{(\alpha', \beta') : (\alpha', \beta') \text{ is well-labelled}, \rho(\alpha') = \alpha, \rho(\beta') = \beta\}.$$

Finally, the labelled product of two classes $\mathcal{A}$ and $\mathcal{B}$ is

$$\mathcal{A} \star \mathcal{B} = \bigcup_{\alpha \in \mathcal{A}, \beta \in \mathcal{B}} (\alpha \star \beta).$$

The EGF can be derived by noting that for objects of size $k$and $n-k$, there are $\binom{n}{k}$ ways to do the relabelling. Therefore, the total number of objects of size $n$is

$$\sum_{k=0}^{n}\binom{n}{k}A_k B_{n-k}.$$

This *binomial convolution* relation for the terms is equivalent to multiplying the EGFs,

$$A(z)\cdot B(z).$$

Set

In labelled structures, a set of $k$elements corresponds to exactly $k!$ sequences. This is different from the unlabelled case, where some of the permutations may coincide. Thus for $\mathcal{A}=\mathfrak{P}\{\mathcal{B}\}$, we have

$$A(z)=\sum_{k=0}^{\infty}\frac{B(z)^k}{k!}=\exp(B(z))$$

Cycle

Cycles are also easier than in the unlabelled case. A cycle of length $k$corresponds to $k$distinct sequences. Thus for $\mathcal{A}=\mathfrak{C}\{\mathcal{B}\}$, we have

$$A(z)=\sum_{k=0}^{\infty}\frac{B(z)^k}{k}=\ln\left(\frac{1}{1-B(z)}\right).$$

Other Elementary Constructions

The operators

$$\mathfrak{C}_{\text{even}},\mathfrak{C}_{\text{odd}},\mathfrak{P}_{\text{even}},\text{ and }\mathfrak{P}_{\text{odd}}$$

represent cycles of even and odd length, and sets of even and odd cardinality.

Example

Stirling numbers of the second kind may be derived and analyzed using the structural decomposition

$$\mathfrak{P}(\mathfrak{P}_{\geq 1}(\mathcal{Z})).$$

The decomposition

$$\mathfrak{P}(\mathfrak{C}(\mathcal{Z}))$$

is used to study unsigned Stirling numbers of the first kind, and in the derivation of the statistics of random permutations. A detailed examination of the exponential generating functions associated to Stirling numbers within symbolic combinatorics may be found on the page on Stirling numbers and exponential generating functions in symbolic combinatorics.

Partition (Number Theory)

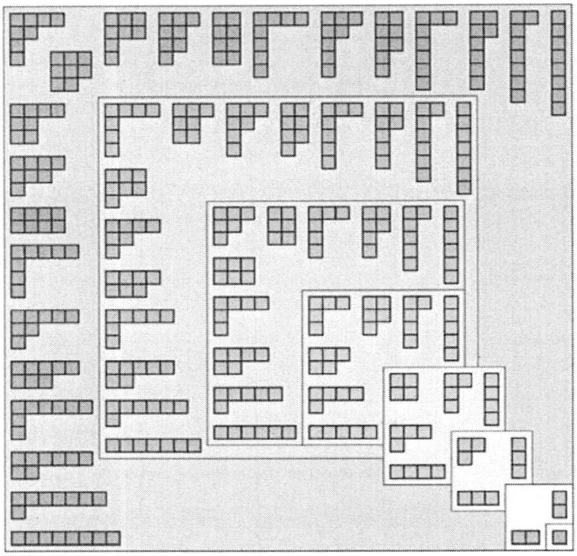

Figure: *Young diagrams associated to the partitions of the positive integers 1 through 8. They are arranged so that images under the reflection about the main diagonal of the square are conjugate partitions.*

In number theory and combinatorics, a partition of a positive integer n, also called an integer partition, is a way of writing n as a sum of positive integers.

Two sums that differ only in the order of their summands are considered to be the same partition; if order matters then the sum becomes a composition.

For example, 4 can be partitioned in five distinct ways:

$$4, \quad 3 + 1, \quad 2 + 2, \quad 2 + 1 + 1, \quad 1 + 1 + 1 + 1.$$

The order-dependent composition 1 + 3 is the same partition as 3 + 1, while 1 + 2 + 1 and 1 + 1 + 2 are the same partition as 2 + 1 + 1.

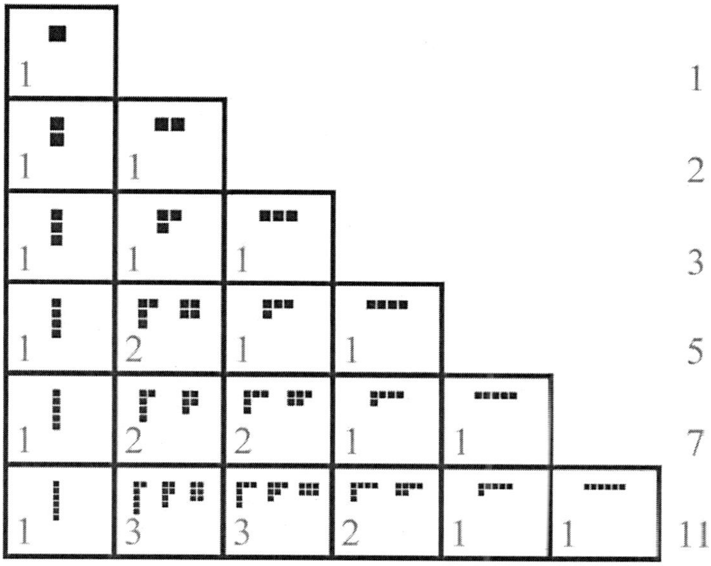

***Figure:** Partitions of* n *with biggest addend* k

A summand in a partition is also called a part. The number of partitions of n is given by the partition function $p(n)$. So $p(4)$ = 5. The notation λ ¢" n means that λ is a partition of n.

Partitions can be graphically visualized with Young diagrams or Ferrers diagrams.

They occur in a number of branches of mathematics and physics, including the study of symmetric polynomials, the symmetric group and in group representation theory in general.

Examples

The partitions of 4 are:

1. 4
2. 3 + 1
3. 2 + 2
4. 2 + 1 + 1
5. 1 + 1 + 1 + 1.

In some sources partitions are treated as the sequence of summands, rather than as an expression with plus signs.

For example, the partition 2 + 1 + 1 might instead be written as the tuple (2, 1, 1) or in the even more compact form $(2, 1^2)$ where the superscript indicates the number of repetitions of a term.

Restricted Partitions

A restricted partition is a partition in which the parts are constrained in some way.

For example, we could count partitions which contain only odd numbers. Among the 22 partitions of the number 8, there are 6 that contain only *odd parts*:

- 7 + 1
- 5 + 3
- 5 + 1 + 1 + 1
- 3 + 3 + 1 + 1
- 3 + 1 + 1 + 1 + 1 + 1
- 1 + 1 + 1 + 1 + 1 + 1 + 1 + 1

Alternatively, we could count partitions in which no number occurs more than once. If we count the partitions of 8 with *distinct parts*, we also obtain 6:

- 8
- 7 + 1
- 6 + 2
- 5 + 3
- 5 + 2 + 1
- 4 + 3 + 1

For all positive numbers the number of partitions with odd parts equals the number of partitions with distinct parts. This result was proved by Leonhard Euler in 1748 and is a special case of Glaisher's theorem. Some similar results about restricted partitions can be obtained by the aid of a visual tool, a Ferrers graph (also called Ferrers diagram, since it is not a *graph* in the graph-theoretical sense, or sometimes Young diagram, alluding to the Young tableau).

Some results concerning restricted partitions are:

- The number of partitions of n in which the greatest part is m is equal to the number of partitions of n into m parts.
- The number of partitions of n in which each part is less than or equal to m is equal to the number of partitions of n into m or fewer parts.
- The number of partitions of n in which all parts are equal is the number of divisors of n.
- The number of partitions of n in which all parts are 1 or 2 (or, equivalently, the number of partitions of n into 1 or 2 parts) is

$$\left\lfloor \frac{n}{2} + 1 \right\rfloor.$$

- The number of partitions of n in which all parts are 1, 2 or 3 (or, equivalently, the number of partitions of n into 1, 2 or 3 parts) is the nearest integer to $(n + 3)^2$ / 12.

Partition Function

In number theory, the partition function $p(n)$ represents the number of possible partitions of a natural number n, which is to say the number of distinct ways of representing n as a sum of natural numbers (with order irrelevant). By convention $p(0) = 1$, $p(n) = 0$ for n negative.

The first few values of the partition function are (starting with $p(0)=1$):

1, 1, 2, 3, 5, 7, 11, 15, 22, 30, 42, ... (sequence A000041 in OEIS).

The value of $p(n)$ has been computed for large values of n, for example $p(100)=190,569,292$ and $p(1000)$ is approximately 2.4×10^{31}.

As of June 2012, the largest known prime number that counts a number of partitions is $p(82352631)$, with 10101 decimal digits.

For every type of restricted partition there is a corresponding function for the number of partitions satisfying the given restriction. An important example is $q(n)$, the number of partitions of n into distinct parts.

As noted above, $q(n)$ is also the number of partitions of n into odd parts. The first few values of $q(n)$ are (starting with $q(0)=1$):

1, 1, 1, 2, 2, 3, 4, 5, 6, 8, 10, ... (sequence A000009 in OEIS).

Generating Function

The generating function for $p(n)$ is given by:

$$\sum_{n=0}^{\infty} p(n)x^n = \prod_{k=1}^{\infty}\left(\frac{1}{1-x^k}\right).$$

Expanding each term on the right-hand side as a geometric series, we can rewrite it as

$(1 + x + x^2 + x^3 + ...)(1 + x^2 + x^4 + x^6 + ...)(1 + x^3 + x^6 + x^9 + ...) ....$

The x^n term in this product counts the number of ways to write

$n = a_1 + 2a_2 + 3a_3 + ... = (1 + 1 + ... + 1) + (2 + 2 + ... + 2) + (3 + 3 + ... + 3) + ...,$

where each number i appears a_i times. This is precisely the definition of a partition of n, so our product is the desired generating function. More generally, the generating function for the partitions of n into numbers from a set A can be found by taking only those terms in the product where k is an element of A. This result is due to Euler.

The formulation of Euler's generating function is a special case of a q-Pochhammer symbol and is similar to the product formulation of many modular forms, and specifically the Dedekind eta function.

The denominator of the product is Euler's function and can be written, by the pentagonal number theorem, as

$$(1-x)(1-x^2)(1-x^3)\ldots = 1-x-x^2+x^5+x^7-x^{12}-x^{15}+x^{22}+x^{26}+\ldots.$$

where the exponents of x on the right hand side are the generalized pentagonal numbers; i.e., numbers of the form $\frac{1}{2}m(3m-1)$, where m is an integer. The signs in the summation alternate as $(-1)^m$. This theorem can be used to derive a recurrence for the partition function:

$p(k) = p(k-1) + p(k-2) - p(k-5) - p(k-7) + p(k-12) + p(k-15) - p(k-22) - \ldots$

where p(0) is taken to equal 1, and p(k) is taken to be zero for negative k.

Another way of stating this is that the value of $p(n)$ can be found from the formula

$$
\begin{array}{l}
\text{GPN's} \\
0 \\
1 \\
2 \\
\\
\\
5 \\
\\
7 \\
\\
\\
\vdots
\end{array}
\qquad
p(n) = \begin{vmatrix}
1 & -1 & & & & & & & & \\
1 & 1 & -1 & & & & & & & \\
0 & 1 & 1 & -1 & & & & & & \\
0 & 0 & 1 & 1 & -1 & & & & & \\
-1 & 0 & 0 & 1 & 1 & -1 & & & & \\
0 & -1 & 0 & 0 & 1 & 1 & -1 & & & \\
-1 & 0 & -1 & 0 & 0 & 1 & 1 & -1 & & \\
0 & -1 & 0 & -1 & 0 & 0 & 1 & 1 & -1 & \\
0 & 0 & -1 & 0 & -1 & 0 & 0 & 1 & 1 & \\
\vdots & & & & & & & & & \ddots
\end{vmatrix}_{n\times n}.
$$

I.e., $p(n)$ is the determinant of the $n\times n$ truncation of the infinite-dimensional Toeplitz matrix shown above. The only non-zero diagonals of this matrix are those which start on a row labelled by a generalized pentagonal number q_m. (The superdiagonal is taken to start on row "0".) On these diagonals, the matrix element is $(-1)^{m+1}$. This follows from a general formula for the quotients for power series.

The generating function for $q(n)$ (partitions into distinct parts) is given by:

$$\sum_{n=0}^{\infty} q(n)x^n = \prod_{k=1}^{\infty}(1+x^k) = \prod_{k=1}^{\infty}\left(\frac{1}{1-x^{2k-1}}\right).$$

The second product can be written $\Phi(x^2) / \Phi(x)$ where Φ is Euler's function; the pentagonal number theorem can be applied to this as well giving a recurrence for q:

$q(k) = a_k + q(k-1) + q(k-2) - q(k-5) - q(k-7) + q(k-12) + q(k-15) - q(k-22) - \ldots$

where a_k is $(-1)^m$ if $k = 3m^2-m$ for some integer m and is 0 otherwise.

The determinant formula for the quotient of power series can be applied to the expression $\Phi(x^2) / \Phi(x)$ to produce the expression

$$q(n) = \begin{vmatrix} 1 & & & & & & & & 1 \\ -1 & 1 & & & & & & & 0 \\ -1 & -1 & 1 & & & & & & -1 \\ 0 & -1 & -1 & 1 & & & & & 0 \\ 0 & 0 & -1 & -1 & 1 & & & & -1 \\ 1 & 0 & 0 & -1 & -1 & 1 & & & 0 \\ 0 & 1 & 0 & 0 & -1 & -1 & 1 & & 0 \\ 1 & 0 & 1 & 0 & 0 & -1 & -1 & & 0 \\ \vdots & & & & & & & \ddots & \vdots \end{vmatrix}_{(n+1)\times(n+1)},$$

where the diagonals in the first n columns are constants equal to the coefficients in the power series for $\Phi(x)$ and the last column has values a_k given above.

Gaussian Binomial Coefficient

The Gaussian binomial coefficient is related to integer partitions. The Gaussian binomial coefficient is defined as:

$$\binom{k+l}{l}_q = \binom{k+l}{k}_q = \frac{\prod_{j=1}^{k+l}(1+q^j)}{\prod_{j=1}^{k}(1+q^j)\prod_{j=1}^{l}(1+q^j)}.$$

The number of integer partitions that would fit into a k by l rectangle (when expressed as a Ferrers or Young diagram) is denoted by $p(n, k, l)$. The Gaussian binomial coefficient is related to the generating function of $p(n, k, l)$ by the following equality:

$$\sum_{n=0}^{\infty} p(n,k,l)x^n = \binom{k+l}{l}_x.$$

Restricted Partition Generating Functions

The generating function can be adapted to describe restricted partitions. For example, the generating function for integer partitions into distinct parts is:

$$\prod_{n=1}^{\infty}(1+x^n)$$

and the generating function for partitions consisting of particular summands (specified by a set T of natural numbers) is:

$$\prod_{t\in T}(1-x^t)^{-1}.$$

This can be used to solve Change-making problems (where the set T specifies the available coins). Generating functions can be used to prove various identities involving integer partitions quite easily, for example the one mentioned in the Restricted partitions section. The generating function for partitions into odd summands is:

$$\prod_{\substack{n=1\\ n \text{ odd}}}^{\infty}(1-x^n)^{-1} = \frac{1}{(1-x)(1-x^3)(1-x^5)\ldots} = \frac{(1-x^2)(1-x^4)\ldots}{(1-x)(1-x^2)(1-x^3)(1-x^4)(1-x^5)\ldots}$$

$$= \frac{(1-x)(1+x)(1-x^2)(1+x^2)\ldots}{(1-x)(1-x^2)(1-x^3)(1-x^4)(1-x^5)\ldots} = (1+x)(1+x^2)(1+x^3)\ldots$$

which is the generating function for partitions into distinct summands.

Congruences

Srinivasa Ramanujan is credited with discovering that "congruences" in the number of partitions exist for arguments that are integers ending in 4 and 9.

$$p(5k+4) \equiv 0 \pmod{5}$$

For instance, the number of partitions for the integer 4 is 5. For the integer 9, the number of partitions is 30; for 14 there are 135 partitions. This is implied by an identity, also by Ramanujan,

$$\sum_{k=0}^{\infty} p(5k+4)x^k = 5\frac{(x^5)_\infty^5}{(x)_\infty^6}$$

where the series $(x)_\infty$ is defined as

$$(x)_\infty = \prod_{m=1}^{\infty}(1-x^m).$$

He also discovered congruences related to 7 and 11:

$$p(7k+5) \equiv 0 \pmod{7}$$
$$p(11k+6) \equiv 0 \pmod{11}.$$

Since 5, 7, and 11 are consecutive primes, one might think that there would be such a congruence for the next prime 13, $p(13k+a) \equiv 0 \pmod{13}$ for some a. This is, however, false. It can also be shown that there is no congruence of the form $p(bk+a) \equiv 0 \pmod{b}$ for any prime b other than 5, 7, or 11.

In the 1960s, A. O. L. Atkin of the University of Illinois at Chicago discovered additional congruences for small prime moduli. For example:

$$p(11^3 \cdot 13 \cdot k + 237) \equiv 0 \pmod{13}.$$

In 2000, Ken Ono of the University of Wisconsin–Madison proved that there are such congruences for every prime modulus. A few years later Ono, together with Scott Ahlgren of the University of Illinois, proved that there are partition congruences modulo every integer coprime to 6.

Partition Function Formulas

Exact formula: Leonhard Euler's pentagonal number theorem implies the identity

$$p(n) = p(n-1) + p(n-2) - p(n-5) - p(n-7) + \cdots$$

where the numbers 1, 2, 5, 7, ... that appear on the right side of the equation are the generalized pentagonal numbers $g_k = \dfrac{k(3k-1)}{2}$ for nonzero integers k. More formally,

$$p(n) = \sum_k (-1)^{k-1} p\left(n - k(3k-1)/2\right)$$

where the summation is over all nonzero integers k (positive and negative) and $p(m)$ is taken to be 0 if $m < 0$.

Approximation Formulas

Approximation formulas exist that are faster to calculate than the exact formula given above. An asymptotic expression for $p(n)$ is given by

$$p(n) \sim \frac{1}{4n\sqrt{3}} \exp\left(\pi\sqrt{\frac{2n}{3}}\right) \text{ as } n \to \infty.$$

This asymptotic formula was first obtained by G. H. Hardy and Ramanujan in 1918 and independently by J. V. Uspensky in 1920. Considering $p(1000)$, the asymptotic formula gives about 2.4402×10^{31}, reasonably close to the exact answer given above (1.415% larger than the true value).

Hardy and Ramanujan obtained an asymptotic expansion with this approximation as the first term:

$$p(n) = \frac{1}{2\sqrt{2}} \sum_{k=1}^{v} \sqrt{k}\, A_k(n) \frac{d}{dn} \exp\left(\pi \sqrt{\frac{2}{3}} \frac{\sqrt{n - \frac{1}{24}}}{k} \right)$$

where

$$A_k(n) = \sum_{0 \le m < k;\, (m,k)=1} e^{\pi i \left[s(m,k) - \frac{1}{k} 2nm \right]}.$$

Here, the notation $(m, n) = 1$ implies that the sum should occur only over the values of m that are relatively prime to n. The function $s(m, k)$ is a Dedekind sum.

The error after v terms is of the order of the next term, and v may be taken to be of the order of $\sqrt{n}$. As an example, Hardy and Ramanujan showed that $p(200)$ is the nearest integer to the sum of the first $v=5$ terms of the series.

In 1937, Hans Rademacher was able to improve on Hardy and Ramanujan's results by providing a convergent series expression for $p(n)$. It is

$$p(n) = \frac{1}{\pi\sqrt{2}} \sum_{k=1}^{\infty} \sqrt{k}\, A_k(n) \frac{d}{dn} \left(\frac{1}{\sqrt{n - \frac{1}{24}}} \sinh\left[\frac{\pi}{k} \sqrt{\frac{2}{3}\left(n - \frac{1}{24} \right)} \right] \right).$$

The proof of Rademacher's formula involves Ford circles, Farey sequences, modular symmetry and the Dedekind eta function in a central way. It may be shown that the k-th term of Rademacher's series is of the order

$$\exp\left(\pi \sqrt{\frac{2}{3}} \frac{\sqrt{n}}{k} \right),$$

so that the first term gives the Hardy–Ramanujan asymptotic approximation.

Paul Erdõs published an elementary proof of the asymptotic formula for $p(n)$ in 1942.

Asymptotics of Restricted Partitions

The asymptotic expression for $p(n)$ implies that

$$\log p(n) \sim C\sqrt{n} \text{ as } n \to \infty$$

where $$C = \pi\sqrt{\frac{2}{3}}.$$

If A is a set of natural numbers, we let $p_A(n)$ denote the number of partitions of n into elements of A. If A possesses positive natural density α then

$$\log p_A(n) \sim C\sqrt{\alpha n}$$

and conversely if this asymptotic property holds for $p_A(n)$ then A has natural density α. This result was stated, with a sketch of proof, by Erdõs in 1942.

If A is a finite set, this analysis does not apply (the density of a finite set is zero). If A has k elements then

$$p_A(n) = \left(\prod_{a \in A} a^{-1}\right) \cdot \frac{n^{k-1}}{(k-1)!} + O(n^{k-2}).$$

Combinatorial Design

Combinatorial design theory is the part of combinatorial mathematics that deals with the existence, construction and properties of systems of finite sets whose arrangements satisfy generalized concepts of *balance* and/or *symmetry*.

These concepts are not made precise so that a wide range of objects can be thought of as being under the same umbrella. At times this might involve the numerical sizes of set intersections as in block designs, while at other times it could involve the spatial arrangement of entries in an array as in Sudoku grids.

Combinatorial design theory can be applied to the area of design of experiments. Some of the basic theory of combinatorial designs originated in the statistician Ronald Fisher's work on the design of biological experiments. Modern applications are also found in a wide gamut of areas including; Finite geometry, tournament scheduling, lotteries, mathematical biology, algorithm design and analysis, networking, group testing and cryptography.

Example

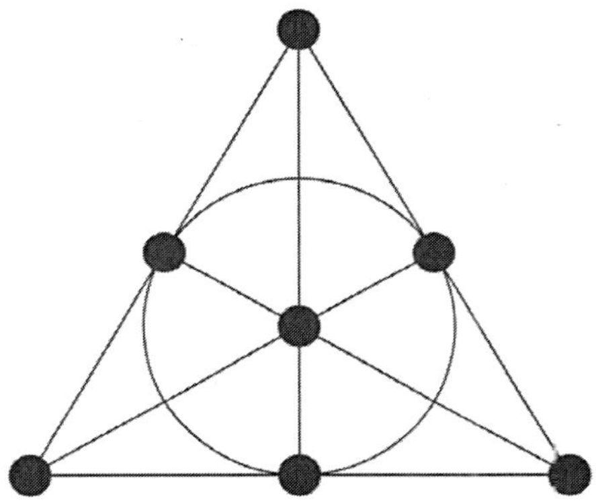

Figure: *The Fano plane*

Given a certain number n of people, is it possible to assign them to sets so that each person is in at least one set, each pair of people is in exactly one set together, every two sets have exactly one person in common, and no set contains everyone, all but one person, or exactly one person? The answer depends on n.

This has a solution only if n has the form $q^2 + q + 1$. It is less simple to prove that a solution exists if q is a prime power. It is conjectured that these are the *only* solutions.

It has been further shown that if a solution exists for q congruent to 1 or 2 mod 4, then q is a sum of two square numbers.

This last result, the Bruck–Ryser theorem, is proved by a combination of constructive methods based on finite fields and an application of quadratic forms.

When such a structure does exist, it is called a finite projective plane; thus showing how finite geometry and combinatorics intersect. When $q = 2$, the projective plane is called the Fano plane.

Fundamental Combinatorial Designs

The classical core of the subject of combinatorial designs is built around balanced incomplete block designs (BIBDs), Hadamard matrices and Hadamard designs, symmetric BIBDs, Latin squares, resolvable BIBDs, difference sets, and pairwise balanced designs (PBDs). Other combinatorial designs are related to or have been developed from the study of these fundamental ones.

- A balanced incomplete block design or BIBD (usually called for short a block design) is a collection B of b subsets (called *blocks*) of a finite set X of v elements, such that any element of X is contained in the same number r of blocks, every block has the same number k of elements, and each pair of distinct elements appear together in the same number λ of blocks. BIBDs are also known as *2-designs* and are often denoted as 2-(v,k,λ) designs. As an example, when $\lambda = 1$ and $b = v$, we have a projective plane: X is the point set of the plane and the blocks are the lines.
- A symmetric balanced incomplete block design or SBIBD is a BIBD in which $v = b$ (the number of points equals the number of blocks). They are the single most important and well studied subclass of BIBDs. Projective planes, biplanes and Hadamard 2-designs are all SBIBDs. They are of particular interest since they are the extremal examples of Fisher's inequality (b eŠ v).
- A resolvable BIBD is a BIBD whose blocks can be partitioned into sets (called *parallel classes*), each of which forms a partition of the point set of the BIBD. The set of parallel classes is called a *resolution* of the design. A solution of the famous 15 schoolgirl problem is a resolution of a BIBD with $v = 15$, $k = 3$ and $\lambda = 1$.
- A Latin rectangle is an $r \times n$ matrix that has the numbers 1, 2, 3, ..., n as its entries (or any other set of n distinct symbols) with no number occurring more than once in any row or column where $r \leq n$. An $n \times n$ Latin rectangle is called a Latin square. If $r < n$, then it is possible to append n ” r rows to an $r \times n$ Latin

rectangle to form a Latin square, using Hall's marriage theorem.

Two Latin squares of order n are said to be *orthogonal* if the set of all ordered pairs consisting of the corresponding entries in the two squares has n^2 distinct members (all possible ordered pairs occur). A set of Latin squares of the same order forms a set of mutually orthogonal Latin squares (MOLS) if every pair of Latin squares in the set are orthogonal. There can be at most n - 1 squares in a set of MOLS of order n. A set of MOLS of order n of maximal size can be used to construct a projective plane of order n.

- A (v, k, λ) difference set is a subset D of a group G such that the order of G is v, the size of D is k, and every nonidentity element of G can be expressed as a product $d_1d_2^{Š1}$ of elements of D in exactly λ ways (when G is written with a multiplicative operation).

If D is a difference set, and g in G, then g D = {gd: d in D} is also a difference set, and is called a translate of D. The set of all translates of a difference set D forms a symmetric block design. In such a design there are v elements and v blocks. Each block of the design consists of k points, each point is contained in k blocks. Any two blocks have exactly λ elements in common and any two points appear together in λ blocks. This SBIBD is called the *development* of D.

In particular, if λ = 1, then the difference set gives rise to a projective plane. An example of a (7,3,1) difference set in the group $\mathbb{Z}/7\mathbb{Z}$ (an abelian group written additively) is the subset {1,2,4}. The development of this difference set gives the Fano plane.

Since every difference set gives an SBIBD, the parameter set must satisfy the Bruck–Ryser–Chowla theorem. but not every SBIBD gives a difference set.

- An Hadamard matrix of order m is an $m \times m$ matrix H whose entries are ±1 such that $HH^{\top} = mI_m$, where $H^{\top}$ is the transpose of H and I_m is the $m \times m$ identity matrix. An Hadamard matrix can be put into *standardized form* (that is, converted to an equivalent Hadamard matrix) where the first row and first column entries are all +1. If the order $m > 2$ then m must be a multiple of 4.

Given an Hadamard matrix of order $4a$ in standardized form, remove the first row and first column and convert every -1 to a 0. The resulting 0-1 matrix M is the incidence matrix of a symmetric 2-($4a$-1, $2a$-1, a-1) design called an Hadamard 2-design.

This construction is reversible, and the incidence matrix of a symmetric 2-design with these parameters can be used to form an Hadamard matrix of order $4a$.

When $a = 2$ we obtain the, by now familiar, Fano plane as an Hadamard 2-design.

- A pairwise balanced design (or PBD) is a set X together with a family of subsets of X (which need not have the same size and may contain repeats) such that every pair of distinct elements of X is contained in exactly λ (a positive integer) subsets. The set X is allowed to be one of the subsets, and if all the subsets are copies of X, the PBD is called *trivial*. The size of X is v and the number of subsets in the family (counted with multiplicity) is b.

Fisher's inequality holds for PBDs: For any non-trivial PBD, $v \leq b$.

This result also generalizes the famous Erdõs–De Bruijn theorem: For a PBD with $\lambda = 1$ having no blocks of size 1 or size v, $v \leq b$, with equality if and only if the PBD is a projective plane or a near-pencil.

A wide Assortment of Other Combinatorial Designs

The Handbook of Combinatorial Designs (Colbourn & Dinitz 2007) has, amongst others, 65 chapters, each devoted to a combinatorial design other than those given above. A partial listing is given below:

- Association schemes
- A balanced ternary design BTD(V, B; ρ_1, ρ_2, R; K, Λ) is an arrangement of V elements into B multisets (blocks), each of cardinality K ($K \leq V$), satisfying:

1. Each element appears $R = \rho_1 + 2\rho_2$ times altogether, with multiplicity one in exactly ρ_1 blocks and multiplicity two in exactly ρ_2 blocks.
2. Every pair of distinct elements appears Λ times (counted with multiplicity); that is, if m_{vb} is the multiplicity of the element v in block b, then for every pair of distinct elements v and w, $\sum_{b=1}^{B} m_{vb}m_{wb} = \Lambda$.

For example, one of the only two nonisomorphic BTD(4,8;2,3,8;4,6)s (blocks are columns) is:

1	1	1	2	2	3	1	1
1	1	1	2	2	3	2	2
2	3	4	3	4	4	3	3
2	3	4	3	4	4	4	4

The incidence matrix of a BTD (where the entries are the multiplicities of the elements in the blocks) can be used to form a ternary error-correcting code analogous to the way binary codes are formed from the incidence matrices of BIBDs.

- A balanced tournament design of order n (a BTD(n)) is an arrangement of all the distinct unordered pairs of a $2n$-set V into an $n \times (2n-1)$ array such that

1. every element of V appears precisely once in each column, and
2. every element of V appears at most twice in each row.

An example of a BTD(3) is given by

1 6	3 5	2 3	4 5	2 4
2 5	4 6	1 4	1 3	3 6
3 4	1 2	5 6	2 6	1 5

The columns of a BTD(n) provide a 1-factorization of the complete graph on $2n$ vertices, K_{2n}.

BTD(n)s can be used to schedule round robin tournaments: the rows represent the locations, the columns the rounds of play and the entries are the competing players or teams.

- Bent functions
- Costas arrays
- Factorial designs
- A frequency square (F-square) is a higher order generalization of a Latin square. Let $S = \{s_1, s_2, ..., s_m\}$ be a set of distinct symbols and $(\lambda_1, \lambda_2, ..., \lambda_m)$ a *frequency vector* of positive integers. A *frequency square* of order n is an $n \times n$ array in which each symbol s_i occurs λ_i times, $i = 1,2,...,m$, in each row and column. The *order* $n = \lambda_1 + \lambda_2 + ... + \lambda_m$. An F-square is in *standard form* if in the first row and column, all occurrences of s_i precede those of s_j whenever $i < j$.

A frequency square F_1 of order n based on the set $\{s_1, s_2, ..., s_m\}$ with frequency vector $(\lambda_1, \lambda_2, ..., \lambda_m)$ and a frequency square F_2, also of order n, based on the set $\{t_1, t_2, ..., t_k\}$ with frequency vector $(\mu_1, \mu_2, ..., \mu_k)$ are *orthogonal* if every ordered pair (s_i, t_j) appears precisely $\lambda_i\mu_j$ times when F_1 and F_2 are superimposed.

- Hall triple systems (HTSs) are Steiner triple systems (STSs) (but the blocks are called *lines*) with the property that the substructure generated by two intersecting lines is isomorphic to the finite affine plane AG(2,3).

Any affine space AG(n,3) gives an example of an HTS. Such an HTS is an *affine* HTS. Nonaffine HTSs also exist.

The number of points of an HTS is 3^m for some integer $m \geq 2$. Nonaffine HTSs exist for any $m \geq 4$ and do not exist for $m = 2$ or 3.

Every Steiner triple system is equivalent to a Steiner quasigroup (idempotent, commutative and satisfying $(xy)y = x$ for all x and y). A Hall triple system is equivalent to a Steiner quasigroup which is distributive, that is, satisfies $a(xy) = (ax)(ay)$ for all a,x,y in the quasigroup.

- Let S be a set of $2n$ elements. A Howell design, $H(s,2n)$ (on symbol set S) is an $s \times s$ array such that:
  1. Each cell of the array is either empty or contains an unordered pair from S,
  2. Each symbol occurs exactly once in each row and column of the array, and
  3. Every unordered pair of symbols occurs in at most one cell of the array.

An example of an H(4,6) is

0 4		1 3	2 5
2 3	1 4	0 5	
	3 5	2 4	0 1
1 5	0 2		3 4

An $H(2n - 1, 2n)$ is a Room square of side $2n - 1$, and thus the Howell designs generalize the concept of Room squares.

The pairs of symbols in the cells of a Howell design can be thought of as the edges of an s regular graph on $2n$ vertices, called the *underlying graph* of the Howell design.

Cyclic Howell designs are used as *Howell movements* in duplicate bridge tournaments. The rows of the design represent the rounds, the columns represent the boards, and the diagonals represent the tables.

- Linear spaces
- An (n,k,p,t)-lotto design is an n-set V of elements together with a set β of k-element subsets of V (blocks), so that for any p-subset P of V, there is a block B in β for which $|P \cap B| \geq t$. $L(n,k,p,t)$ denotes the smallest number of blocks in any (n,k,p,t)-lotto design. The following is a (7,5,4,3)-lotto design with the smallest possible number of blocks:

  $\{1,2,3,4,7\}$ $\{1,2,5,6,7\}$ $\{3,4,5,6,7\}$.

Lotto designs model any lottery that is run in the following way: Individuals purchase *tickets* consisting of k numbers chosen from a set of n numbers. At a certain point the sale of tickets is stopped and a set of p numbers is randomly selected from the n numbers. These are the *winning numbers*. If any sold ticket contains t or more of the winning numbers, a prize is given to the ticket holder. Larger prizes go to tickets with more matches. The value of L(n,k,p,t) is of interest to both gamblers and researchers, as this is the smallest number of tickets that are needed to be purchased in order to guarantee a prize.

The Hungarian Lottery is a (90,5,5,t)-lotto design and it is known that L(90,5,5,2) = 100. Lotteries with parameters (49,6,6,t) are also popular worldwide and it is known that L(49,6,6,2) = 19. In general though, these numbers are hard to calculate and remain unknown.

A geometric construction of one such design is given in Transylvanian lottery.

- Magic squares
- A (v,k,λ)-Mendelsohn design, or MD(v,k,λ),is a v-set V and a collection β of ordered k-tuples of distinct elements of V (called *blocks*), such that each ordered pair (x,y) with $x \neq y$ of elements of V is cyclically adjacent in λ blocks. The ordered pair (x,y) of distinct elements is *cyclically adjacent* in a block if the elements appear in the block as (...,x,y,...) or (y,...,x). An MD(v,3,λ) is a Mendelsohn triple system, MTS(v,λ). An example of an MTS(4,1) on V = {0,1,2,3} is:

(0,1,2) (1,0,3) (2,1,3) (0,2,3)

Any triple system can be made into a Mendelson triple system by replacing the unordered triple {a,b,c} with the pair of ordered triples (a,b,c) and (a,c,b), but as the example shows, the converse of this statement is not true.

If (Q,*) is an idempotent semisymmetric quasigroup, that is, $x * x = x$ (idempotent) and $x * (y * x) = x$ (semisymmetric) for all x, y in Q, let β = {(x,y,$x * y$): x, y in Q}. Then (Q, β) is a Mendelsohn triple system MTS($|Q|$,1). This construction is reversible.

- Orthogonal arrays
- A quasi-3 design is a symmetric design (SBIBD) in which each triple of blocks intersect in either x or y

points, for fixed x and y called the *triple intersection numbers* ($x < y$). Any symmetric design with $\lambda \leq 2$ is a quasi-3 design with $x = 0$ and $y = 1$. The point-hyperplane design of PG(n,q) is a quasi-3 design with $x = (q^{n-2} - 1)/(q - 1)$ and $y = \lambda = (q^{n-1} - 1)/(q - 1)$. If $y = \lambda$ for a quasi-3 design, the design is isomorphic to PG(n,q) or a projective plane.

- A t-(v,k,λ) design D is quasi-symmetric with intersection numbers x and y ($x < y$) if every two distinct blocks intersect in either x or y points. These designs naturally arise in the investigation of the duals of designs with $\lambda = 1$. A non-symmetric ($b > v$) 2-(v,k,1) design is quasisymmetric with $x = 0$ and $y = 1$. A multiple (repeat all blocks a certain number of times) of a symmetric 2-(v,k,λ) design is quasisymmetric with $x = \lambda$ and $y = k$. Hadamard 3-designs (extensions of Hadamard 2-designs) are quasisymmetric.

Every quasisymmetric block design gives rise to a strongly regular graph (as its block graph), but not all SRGs arise in this way. The incidence matrix of a quasisymmetric 2-(v,k,λ) design with $k \equiv x \equiv y$ (mod 2) generates a binary self-orthogonal code (when bordered if k is odd).

- Room squares
- A spherical design is a finite set X of points in a (d – 1)-dimensional sphere such that, for some integer t, the average value on X of every polynomial

$$f(x_1,\ldots,x_d)$$

of total degree at most t is equal to the average value of f on the whole sphere, i.e., the integral of f divided by the area of the sphere.

- Turán systems
- An $r \times n$ tuscan-k rectangle on n symbols has r rows and n columns such that:

1. each row is a permutation of the n symbols and
2. for any two distinct symbols a and b and for each m from 1 to k, there is at most one row in which b is m steps to the right of a.

If $r = n$ and $k = 1$ these are referred to as Tuscan squares, while if $r = n$ and $k = n - 1$ they are Florentine squares. A Roman square is a tuscan square which is also a latin square (these are also known as *row complete latin squares*). A Vatican square is a florentine square which is also a latin square.

The following example is a tuscan-1 square on 7 symbols which is not tuscan-2:

6	1	5	2	4	3	7
2	6	3	5	4	7	1
5	7	2	3	1	4	6
4	2	5	1	6	7	3
3	6	2	1	7	4	5
1	3	2	7	5	6	4
7	6	5	3	4	1	2

A tuscan square on n symbols is equivalent to a decomposition of the complete graph with n vertices into n hamiltonian directed paths.

In a sequence of visual impressions, one flash card may have some effect on the impression given by the next. This bias can be cancelled by using n sequences corresponding to the rows of an $n \times n$ tuscan-1 square.

- A t-wise balanced design (or t BD) of type t-(v,K,λ) is a v-set X together with a family of subsets of X (called *blocks*) whose sizes are in the set K, such that every t-subset of distinct elements of X is contained in exactly λ blocks. If K is a set of positive integers strictly between t and v, then the t BD is *proper*. If all the k-subsets of X for some k are blocks, the t BD is a *trivial design*.

Notice that in the following example of a 3-{12,{4,6},1) design based on the set $X = \{1,2,...,12\}$, some pairs appear four times (such as 1,2) while others appear five times (6,12 for instance).

1 2 3 4 5 6	1 2 7 8	1 2 9 11	1 2 10 12	3 5 7 8	3 5 9 11	3 5 10 12
	4 6 7 8	4 6 9 11	4 6 10 12			
7 8 9 10 11 12	2 3 8 9	2 3 10 7	2 3 11 12	4 1 8 9	4 1 10 7	4 1 11 12
	5 6 8 9	5 6 10 7	5 6 11 12			
	3 4 9 10	3 4 11 8	3 4 7 12	5 2 9 10	5 2 11 8	5 2 7 12
	1 6 9 10	1 6 11 8	1 6 7 12			
	4 5 10 11	4 5 7 9	4 5 8 12	1 3 10 11	1 3 7 9	1 3 8 12
	2 6 10 11	2 6 7 9	2 6 8 12			
	5 1 11 7	5 1 8 10	5 1 9 12	2 4 11 7	2 4 8 10	2 4 9 12
	3 6 11 7	3 6 8 10	3 6 9 12			

- A Youden square is a $k \times v$ *rectangular* array ($k < v$) of v symbols such that each symbol appears exactly once in each row and the symbols appearing in any column form a block of a symmetric (v, k, λ) design, all the blocks of which occur in this manner. A Youden square is a Latin rectangle. The term "square" in the name comes from an older definition which did use a square array. An example of a 4 × 7 Youden square is given by:

1	2	3	4	5	6	7
2	3	4	5	6	7	1
3	4	5	6	7	1	2
5	6	7	1	2	3	4

The seven blocks (columns) form the order 2 biplane (a symmetric (7,4,2)-design).

Finite Geometry

Finite geometry is the study of geometric systems having only a finite number of points. Structures analogous to those found in continuous geometries (Euclidean plane, real projective space, etc.) but defined combinatorially are the main items studied. This area provides a rich source of examples for Design theory. It should not be confused with Discrete geometry (Combinatorial geometry).

Order Theory

Order theory is the study of partially ordered sets, both finite and infinite. Various examples of partial orders appear in algebra, geometry, number theory and throughout combinatorics and graph theory. Notable classes and examples of partial orders include lattices and Boolean algebras.

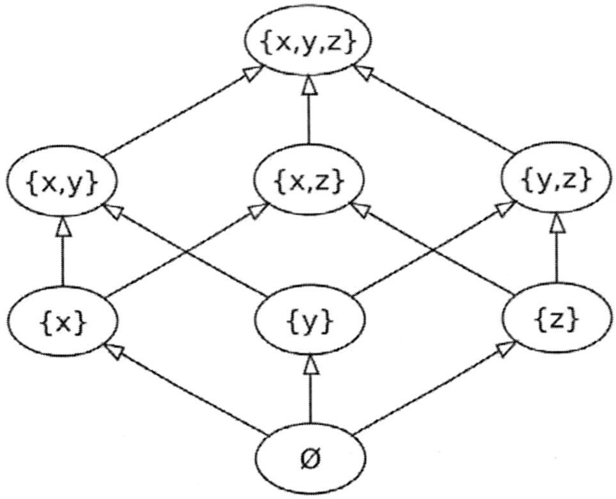

Figure: *Hasse diagram of the powerset of {x,y,z} ordered by inclusion.*

Matroid Theory

Matroid theory abstracts part of geometry. It studies the properties of sets (usually, finite sets) of vectors in a vector space that do not depend on the particular coefficients in a linear dependence relation. Not only the structure but also enumerative properties belong to matroid theory. Matroid theory was introduced by Hassler Whitney and studied as a part of the order theory. It is now an independent field of study with a number of connections with other parts of combinatorics.

Extremal Combinatorics

Extremal combinatorics studies extremal questions on set systems. The types of questions addressed in this case are about the largest possible graph which satisfies certain properties. For example, the largest triangle-free graph on $2n$ vertices is a complete bipartite graph $K_{n,n}$. Often it is too hard even to find the extremal answer $f(n)$ exactly and one can only give an asymptotic estimate.

Ramsey theory is another part of extremal combinatorics. It states that any sufficiently large configuration will contain some sort of order. It is an advanced generalization of the pigeonhole principle.

Probabilistic Combinatorics

Figure: *Self-avoiding walk in a square grid graph.*

In probabilistic combinatorics, the questions are of the following type: what is the probability of a certain property for a random discrete object, such as a random graph? For instance, what is the average number of triangles in a random graph? Probabilistic methods are also used to determine the existence of combinatorial objects with certain prescribed properties (for which explicit examples might be difficult to find), simply by observing that the probability of randomly selecting an object with those properties is greater than 0.

This approach (often referred to as *the* probabilistic method) proved highly effective in applications to extremal combinatorics and graph theory. A closely related area is the study of finite Markov chains, especially on combinatorial objects. Here again probabilistic tools are used to estimate the mixing time.

Often associated with Paul Erdõs, who did the pioneer work on the subject, probabilistic combinatorics was traditionally viewed as a set of tools to study problems in other parts of combinatorics.

However, with the growth of applications to analysis of algorithms in computer science, as well as classical probability, additive and probabilistic number theory, the area recently grew to become an independent field of combinatorics.

Algebraic Combinatorics

Algebraic combinatorics is an area of mathematics that employs methods of abstract algebra, notably group theory and representation theory, in various combinatorial contexts and, conversely, applies combinatorial techniques to problems in algebra.

Within the last decade or so, algebraic combinatorics came to be seen more expansively as the area of mathematics where the interaction of combinatorial and algebraic methods is particularly strong and significant. One of the fastest developing subfields within algebraic combinatorics is combinatorial commutative algebra.

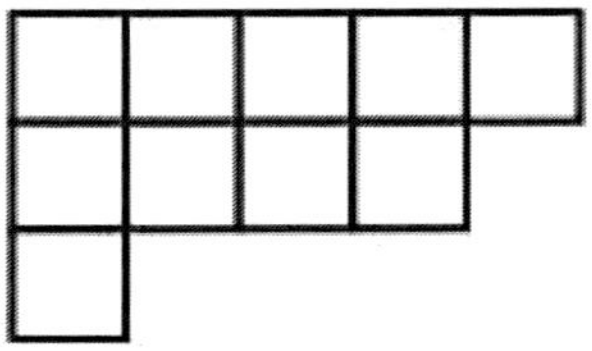

Figure: *Young diagram of a partition (5,4,1).*

Combinatorics on Words

Combinatorics on words deals with formal languages. It arose independently within several branches of mathematics, including number theory, group theory and probability. It has applications to enumerative combinatorics, fractal analysis, theoretical computer science, automata theory and linguistics. While many applications are new, the classical Chomsky–Schützenberger hierarchy of classes of formal grammars is perhaps the best known result in the field.

Geometric Combinatorics

Geometric combinatorics is related to convex and discrete geometry, in particular polyhedral combinatorics. It asks, for example, how many faces of each dimension can a convex polytope have. Metric properties of polytopes play an important role as well, e.g. the Cauchy theorem on rigidity of convex polytopes. Special polytopes are also considered, such as permutohedra, associahedra and Birkhoff polytopes. We should note that combinatorial geometry is an old fashioned name for discrete geometry.

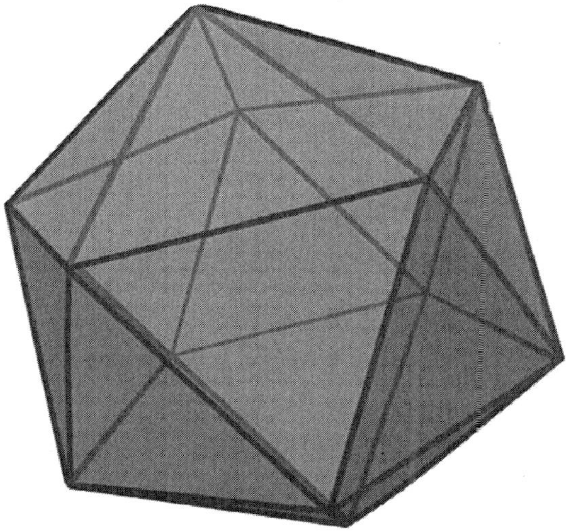

Figure: *An icosahedron.*

Topological Combinatorics

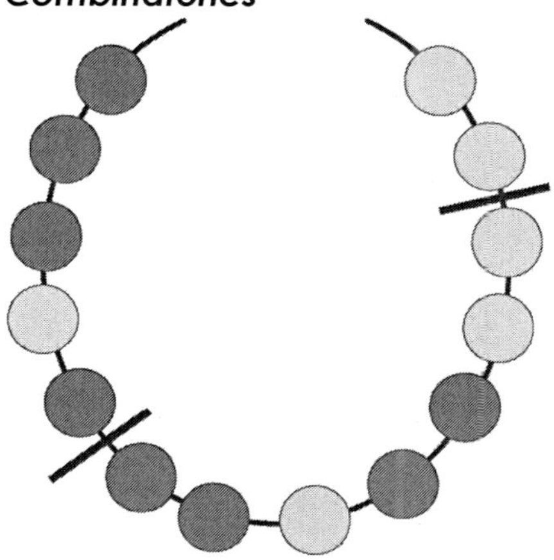

Figure: *Splitting a necklace with two cuts.*

Combinatorial analogs of concepts and methods in topology are used to study graph colouring, fair division, partitions, partially ordered sets, decision trees, necklace problems and discrete Morse theory. It should not be confused with combinatorial topology which is an older name for algebraic topology.

Arithmetic Combinatorics

Arithmetic combinatorics arose out of the interplay between number theory, combinatorics, ergodic theory and harmonic analysis. It is about combinatorial estimates associated with arithmetic operations (addition, subtraction, multiplication, and division). *Additive combinatorics* refers to the special case when only the operations of addition and subtraction are involved. One important technique in arithmetic combinatorics is the ergodic theory of dynamical systems.

Infinitary Combinatorics

Infinitary combinatorics, or combinatorial set theory, is an extension of ideas in combinatorics to infinite sets. It is a part of set theory, an area of mathematical logic, but uses tools and ideas from both set theory and extremal combinatorics.

Gian-Carlo Rota used the name *continuous combinatorics* to describe probability and measure theory, since there are many analogies between *counting* and *measure.*

Related Fields

Figure: *Kissing spheres are connected to both coding theory and discrete geometry.*

Combinatorial Optimization

Combinatorial optimization is the study of optimization on discrete and combinatorial objects. It started as a part of combinatorics and graph theory, but is now viewed as a branch of applied mathematics and computer science, related to operations research, algorithm theory and computational complexity theory.

Coding Theory

Coding theory started as a part of design theory with early combinatorial constructions of error-correcting codes. The main idea of the subject is to design efficient and reliable methods of data transmission. It is now a large field of study, part of information theory.

Discrete and Computational Geometry

Discrete geometry (also called combinatorial geometry) also began a part of combinatorics, with early results on convex polytopes and kissing numbers. With the emergence of applications of discrete geometry to computational geometry, these two fields partially merged and became a separate field of study. There remain many connections with geometric and topological combinatorics, which themselves can be viewed as outgrowths of the early discrete geometry.

Combinatorics and Dynamical Systems

Combinatorial aspects of dynamical systems is another emerging field. Here dynamical systems can be defined on combinatorial objects.

Combinatorics and Physics

There are increasing interactions between combinatorics and physics, particularly statistical physics. Examples include an exact solution of the Ising model, and a connection between the Potts model on one hand, and the chromatic and Tutte polynomials on the other hand.

Chapter 5

Probability

Probability (or likelihood) is a measure or estimation of how likely it is that something will happen or that a statement is true. Probabilities are given a value between 0 (0% chance or *will not happen*) and 1 (100% chance or *will happen*). The higher the degree of probability, the more likely the event is to happen, or, in a longer series of samples, the greater the number of times such event is expected to happen.

These concepts have been given an axiomatic mathematical derivation in probability theory, which is used widely in such areas of study as mathematics, statistics, finance, gambling, science, artificial intelligence/machine learning and philosophy to, for example, draw inferences about the expected frequency of events. Probability theory is also used to describe the underlying mechanics and regularities of complex systems.

Probability Interpretations

The word probability has been used in a variety of ways since it was first applied to the mathematical study of games of chance. Does probability measure the real, physical tendency of something to occur or is it a measure of how strongly one believes it will occur, or does it draw on both these elements? In answering such questions, we interpret the probability values of probability theory.

There are two broad categories of probability interpretations which can be called "physical" and "evidential" probabilities. Physical probabilities, which are also called objective or frequency probabilities, are associated with random physical systems such as roulette wheels, rolling dice and radioactive atoms. In such systems, a given type of event (such as the dice yielding a six) tends to occur at a persistent rate, or "relative frequency", in a long run of trials. Physical probabilities either explain, or are invoked to explain, these stable frequencies.

Thus talk about physical probability makes sense only when dealing with well defined random experiments. The two main kinds of theory of physical probability are frequentist accounts (such as those of Venn, Reichenbach and von Mises) and propensity accounts (such as those of Popper, Miller, Giere and Fetzer).

Evidential probability, also called Bayesian probability (or subjectivist probability), can be assigned to any statement whatsoever, even when no random process is involved, as a way to represent its subjective plausibility, or the degree to which the statement is supported by the available evidence. On most accounts, evidential probabilities are considered to be degrees of belief, defined in terms of dispositions to gamble at certain odds. The four main evidential interpretations are the classical (e.g. Laplace's) interpretation, the subjective interpretation (de Finetti and Savage), the epistemic or inductive interpretation (Ramsey, Cox) and the logical interpretation (Keynes and Carnap).

Some interpretations of probability are associated with approaches to statistical inference, including theories of estimation and hypothesis testing. The physical interpretation, for example, is taken by followers of "frequentist" statistical methods, such as R. A. Fisher, Jerzy Neyman and Egon Pearson. Statisticians of the opposing Bayesian school typically accept the existence and importance of physical probabilities, but also consider the calculation of evidential probabilities to be both valid and necessary in statistics. This article, however, focuses on the interpretations of probability rather than theories of statistical inference.

The terminology of this topic is rather confusing, in part because probabilities are studied within a variety of academic fields. The word "frequentist" is especially tricky. To philosophers it refers to a particular theory of physical probability, one that has more or less been abandoned. To scientists, on the other hand, "frequentist probability" is just what philosophers call physical (or objective) probability. Those who promote Bayesian inference view "frequentist statistics" as an approach to statistical inference that recognises only physical probabilities. Also the word "objective", as applied to probability, sometimes means exactly what "physical" means here, but is also used of evidential probabilities that are fixed by rational constraints, such as logical and epistemic probabilities.

It is unanimously agreed that statistics depends somehow on probability. But, as to what probability is and how it is connected with statistics, there has seldom been such complete disagreement and breakdown of communication since the Tower of Babel. Doubtless, much of the disagreement is merely terminological and would disappear under sufficiently sharp analysis.

Philosophy

The philosophy of probability presents problems chiefly in matters of epistemology and the uneasy interface between mathematical concepts and ordinary language as it is used by non-mathematicians. Probability theory is an established field of study in mathematics.

It has its origins in correspondence discussing the mathematics of games of chance between Blaise Pascal and Pierre de Fermat in the seventeenth century, and was formalized and rendered axiomatic as a distinct branch of mathematics by Andrey Kolmogorov in the twentieth century. In its axiomatic form, mathematical statements about probability theory carry the same sort of epistemological confidence shared by other mathematical statements in the philosophy of mathematics.

The mathematical analysis originated in observations of the behaviour of game equipment such as playing cards and dice, which are designed specifically to introduce random and equalized elements; in mathematical terms, they are subjects of indifference.

This is not the only way probabilistic statements are used in ordinary human language: when people say that "*it will probably rain*", they typically do not mean that the outcome of rain versus not-rain is a random factor that the odds currently favour; instead, such statements are perhaps better understood as qualifying their expectation of rain with a degree of confidence.

Likewise, when it is written that "the most probable explanation" of the name of Ludlow, Massachusetts "is that it was named after Roger Ludlow", what is meant here is not that Roger Ludlow is favoured by a random factor, but rather that this is the most plausible explanation of the evidence, which admits other, less likely explanations.

Thomas Bayes attempted to provide a logic that could handle varying degrees of confidence; as such, Bayesian probability is an attempt to recast the representation of probabilistic statements as an expression of the degree of confidence by which the beliefs they express are held.

Though probability initially had somewhat mundane motivations, its modern influence and use is widespread ranging from Evidence based medicine, through Six sigma, all the way to the Probabilistically checkable proof and the String theory landscape.

Classical Definition

The first attempt at mathematical rigour in the field of probability, championed by Pierre-Simon Laplace, is now known as the classical definition. Developed from studies of games of chance (such as rolling dice) it states that probability is shared equally between all the possible outcomes, provided these outcomes can be deemed equally likely.

> *The theory of chance consists in reducing all the events of the same kind to a certain number of cases equally possible, that is to say, to such as we may be equally undecided about in regard to their existence, and in determining the number of cases favourable to the event whose probability is sought. The ratio of this number to that of all the cases possible is the measure of this probability, which is thus simply a fraction whose numerator is the number of favourable cases and whose denominator is the number of all the cases possible.* — Pierre-Simon Laplace, A Philosophical Essay on Probabilities

Figure: *The classical definition of probability works well for situations with only a finite number of equally-likely outcomes.*

This can be represented mathematically as follows: If a random experiment can result in N mutually exclusive and equally likely outcomes and if N_A of these outcomes result in the occurrence of the event A, the probability of A is defined by

$$P(A) = \frac{N_A}{N}.$$

There are two clear limitations to the classical definition. Firstly, it is applicable only to situations in which there is only a 'finite' number of possible outcomes. But some important random experiments, such as tossing a coin until it rises heads, give rise to an infinite set of outcomes. And secondly, you need to determine in advance that all the possible outcomes are equally likely without relying on the notion of probability to avoid circularity—for instance, by symmetry considerations.

Frequentism

Frequentists posit that the probability of an event is its relative frequency over time, i.e., its relative frequency of occurrence after repeating a process a large number of times under similar conditions. This is also known as aleatory probability.

The events are assumed to be governed by some random physical phenomena, which are either phenomena that are predictable, in principle, with sufficient information; or phenomena which are essentially unpredictable. Examples of the first kind include tossing dice or spinning a roulette wheel; an example of the second kind is radioactive decay. In the case of tossing a fair coin, frequentists say that the probability of getting a heads is 1/2, not because there are two equally likely outcomes but because repeated series of large numbers of trials demonstrate that the empirical frequency converges to the limit 1/2 as the number of trials goes to infinity.

Figure: *For frequentists, the probability of the ball landing in any pocket can be determined only by repeated trials in which the observed result converges to the underlying probability* in the long run.

If we denote by n_a the number of occurrences of an event $\mathcal{A}$ in n trials, then if $\lim_{n\to\infty}\frac{n_a}{n}=p$ we say that $P(\mathcal{A})=p$

The frequentist view has its own problems. It is of course impossible to actually perform an infinity of repetitions of a random experiment to determine the probability of an event. But if only a finite number of repetitions of the process are performed, different relative frequencies will appear in different series of trials. If these relative frequencies are to define the probability, the probability will be slightly different every time it is measured. But the real probability should be the same every time. If we acknowledge the fact that we only can measure a probability with some error of measurement attached, we still get into problems as the error of measurement can only be expressed as a probability, the very concept we are trying to define. This renders even the frequency definition circular.

Logical, Epistemic and Inductive Probability

It is widely recognised that the term "probability" is sometimes used in contexts where it has nothing to do with physical randomness. Consider, for example, the claim that the extinction of the dinosaurs was probably caused by a large meteorite hitting the earth.

Statements such as "Hypothesis H is probably true" have been interpreted to mean that the (presently available) empirical evidence (E, say) supports H to a high degree. This degree of support of H by E has been called the logical probability of H given E, or the epistemic probability of H given E, or the inductive probability of H given E.

The differences between these interpretations are rather small, and may seem inconsequential. One of the main points of disagreement lies in the relation between probability and belief. Logical probabilities are conceived (for example in Keynes' Treatise on Probability) to be objective, logical relations between propositions (or sentences), and hence not to depend in any way upon belief. They are degrees of (partial) entailment, or degrees of logical consequence, not degrees of belief. (They do, nevertheless, dictate proper degrees of belief, as is discussed below.)

Frank P. Ramsey, on the other hand, was skeptical about the existence of such objective logical relations and argued that (evidential) probability is "the logic of partial belief" ("Truth and Probability", 1926, p. 157). In other words, Ramsey held that epistemic probabilities simply *are* degrees of rational belief, rather than being logical relations that merely *constrain* degrees of rational belief.

Another point of disagreement concerns the *uniqueness* of evidential probability, relative to a given state of knowledge. Rudolf Carnap held, for example, that logical principles always determine a unique logical probability for any statement, relative to any body of evidence. Ramsey, by contrast, thought that while degrees of belief are subject to some rational constraints (such as, but not limited to, the axioms of probability) these constraints usually do not determine a unique value. Rational people, in other words, may differ somewhat in their degrees of belief, even if they all have the same information.

Propensity

Propensity theorists think of probability as a physical propensity, or disposition, or tendency of a given type of physical situation to yield an outcome of a certain kind or to yield a long run relative frequency of such an outcome. This kind of objective probability is sometimes called 'chance'.

Propensities, or chances, are not relative frequencies, but purported causes of the observed stable relative frequencies. Propensities are invoked to explain why repeating a certain kind of experiment will generate a given outcome type at a persistent rate,which are known as propensities or chances. Frequentists are unable to take this approach, since relative frequencies do not exist for single tosses of a coin, but only for large ensembles or collectives.

In contrast, a propensitist is able to use the law of large numbers to explain the behaviour of long-run frequencies. This law, which is a consequence of the axioms of probability, says that if (for example) a coin is tossed repeatedly many times, in such a way that its probability of landing heads is the same on each toss, and the outcomes are probabilistically independent, then the relative frequency of heads will be close to the probability of heads on each single toss. This law allows that stable long-run frequencies are a manifestation of invariant *single-case* probabilities. In addition to explaining the emergence of stable relative frequencies, the idea of propensity is motivated by the desire to make sense of single-case probability attributions in quantum mechanics, such as the probability of decay of a particular atom at a particular time.

The main challenge facing propensity theories is to say exactly what propensity means. (And then, of course, to show that propensity thus defined has the required properties.) At present, unfortunately, none of the well-recognised accounts of propensity comes close to meeting this challenge.

A propensity theory of probability was given by Charles Sanders Peirce. A later propensity theory was proposed by philosopher Karl Popper, who had only slight acquaintance with the writings of C. S. Peirce, however. Popper noted that the outcome of a physical experiment is produced by a certain set of "generating conditions". When we repeat an experiment, as the saying goes, we really perform another experiment with a (more or less) similar set of generating conditions. To say that a set of generating conditions has propensity p of producing the outcome E means that those exact conditions, if repeated indefinitely, would produce an outcome sequence in which E occurred with limiting relative frequency p.

For Popper then, a deterministic experiment would have propensity 0 or 1 for each outcome, since those generating conditions would have same outcome on each trial. In other words, non-trivial propensities (those that differ from 0 and 1) only exist for genuinely indeterministic experiments.

A number of other philosophers, including David Miller and Donald A. Gillies, have proposed propensity theories somewhat similar to Popper's.

Other propensity theorists (e.g. Ronald Giere) do not explicitly define propensities at all, but rather see propensity as defined by the theoretical role it plays in science. They argue, for example, that physical magnitudes such as electrical charge cannot be explicitly defined either, in terms of more basic things, but only in terms of what they do (such as attracting and repelling other electrical charges). In a similar way, propensity is whatever fills the various roles that physical probability plays in science.

What roles does physical probability play in science? What are its properties? One central property of chance is that, when known, it constrains rational belief to take the same numerical value. David Lewis called this the *Principal Principle*, a term that philosophers have mostly adopted. For example, suppose you are certain that a particular biased coin has propensity 0.32 to land heads every time it is tossed. What is then the correct price for a gamble that pays $1 if the coin lands heads, and nothing otherwise? According to the Principal Principle, the fair price is 32 cents.

Subjectivism

Subjectivists, also known as Bayesians or followers of epistemic probability, give the notion of probability a subjective status by regarding it as a measure of the 'degree of belief' of the individual assessing the uncertainty of a particular situation.

Epistemic or subjective probability is sometimes called credence, as opposed to the term chance for a propensity probability.

Some examples of epistemic probability are to assign a probability to the proposition that a proposed law of physics is true, and to determine how probable it is that a suspect committed a crime, based on the evidence presented.

Gambling odds don't reflect the bookies' belief in a likely winner, so much as the other bettors' belief, because the bettors are actually betting against one another. The odds are set based on how many people have bet on a possible winner, so that even if the high odds players always win, the bookies will always make their percentages anyway.

The use of Bayesian probability raises the philosophical debate as to whether it can contribute valid justifications of belief.

Bayesians point to the work of Ramsey and de Finetti as proving that subjective beliefs must follow the laws of probability if they are to be coherent.

The use of Bayesian probability involves specifying a prior probability. This may be obtained from consideration of whether the required prior probability is greater or lesser than a reference probability associated with an urn model or a thought experiment. The issue is that for a given problem, multiple thought experiments could apply, and choosing one is a matter of judgement: different people may assign different prior probabilities, known as the reference class problem. The "sunrise problem" provides an example.

Practical Controversy

This difference in point of view has also many implications both for the methods by which statistics is practiced, and for the way in which conclusions are expressed. When comparing two hypotheses and using some information, frequency methods would typically result in the rejection or non-rejection of the original hypothesis at a particular significance level, and frequentists would all agree that the hypothesis should be rejected or not at that level of significance. However, there is no normative methodology to choose levels of significance. Bayesian methods would suggest that one hypothesis was more probable than the other, but individual Bayesians might differ about which was the more probable and by how much, by virtue of having used different priors; but that's the same thing as disagreeing on significance levels, except significance levels are just an *ad hoc* device which are not really a probability, while priors are not only justified by the rules of probability, but there is definitely a normative methodology to define beliefs; so even if a Bayesian wanted to express complete ignorance (as a frequentist claims to do but does it wrong), they could do it with the maximum entropy principle.

The most important distinction between the frequentist and Bayesian paradigms, is that frequentist makes strong distinctions between probability, statistics, and decision-making, whereas Bayesians unify decision-making, statistics and probability under a single philosophically and mathematically consistent framework, unlike the frequentist paradigm which has been proven to be inconsistent, especially for real-world situations where experiments (or "random events") can not be repeated more than once.

Bayesians would argue that this is right and proper — if the issue is such that reasonable people can put forward different, but plausible, priors and the data are such that the likelihood does not swamp the prior, then the issue is not resolved unambiguously at the present stage of knowledge and Bayesian statistics highlights this fact. They would argue that any approach that purports to produce a single, definitive answer to the question at hand in these circumstances is obscuring the truth. But "frequentists" do not claim to produce "a single, definitive answer to the question at hand".

An alternative solution, is the eclectic view, which accepts both interpretations: depending on the situation, one selects one of the two interpretations for pragmatic, or principled, reasons.

Prediction

An alternative account of probability emphasizes the role of *prediction* – predicting future observations on the basis of past observations, not on unobservable parameters. In its modern form, it is mainly in the Bayesian vein. This was the main function of probability before the 20th century, but fell out of favour compared to the parametric approach, which modelled phenomena as a physical system that was observed with error, such as in celestial mechanics.

The modern predictive approach was pioneered by Bruno de Finetti, with the central idea of exchangeability – that future observations should behave like past observations.

This view came to the attention of the Anglophone world with the 1974 translation of de Finetti's book, and has since been propounded by such statisticians as Seymour Geisser.

Axiomatic Probability

The mathematics of probability can be developed on an entirely axiomatic basis that is independent of any interpretation.

Probability Theory

Probability theory is the branch of mathematics concerned with probability, the analysis of random phenomena. The central objects of probability theory are random variables, stochastic processes, and events: mathematical abstractions of non-deterministic events or measured quantities that may either be single occurrences or evolve over time in an apparently random fashion.

If an individual coin toss or the roll of dice is considered to be a random event, then if repeated many times the sequence of random events will exhibit certain patterns, which can be studied and predicted. Two representative mathematical results describing such patterns are the law of large numbers and the central limit theorem.

As a mathematical foundation for statistics, probability theory is essential to many human activities that involve quantitative analysis of large sets of data.

Methods of probability theory also apply to descriptions of complex systems given only partial knowledge of their state, as in statistical mechanics.

A great discovery of twentieth century physics was the probabilistic nature of physical phenomena at atomic scales, described in quantum mechanics.

History

The mathematical theory of probability has its roots in attempts to analyze games of chance by Gerolamo Cardano in the sixteenth century, and by Pierre de Fermat and Blaise Pascal in the seventeenth century (for example the "problem of points"). Christiaan Huygens published a book on the subject in 1657 and in the 19th century a big work was done by Laplace in what can be considered today as the classic interpretation.

Initially, probability theory mainly considered discrete events, and its methods were mainly combinatorial. Eventually, analytical considerations compelled the incorporation of continuous variables into the theory.

This culminated in modern probability theory, on foundations laid by Andrey Nikolaevich Kolmogorov. Kolmogorov combined the notion of sample space, introduced by Richard von Mises, and measure theory and presented his axiom system for probability theory in 1933. Fairly quickly this became the mostly undisputed axiomatic basis for modern probability theory but alternatives exist, in particular the adoption of finite rather than countable additivity by Bruno de Finetti.

Treatment

Most introductions to probability theory treat discrete probability distributions and continuous probability distributions separately. The more mathematically advanced measure theory based treatment of probability covers both the discrete, the continuous, any mix of these two and more.

Motivation

Consider an experiment that can produce a number of outcomes. The collection of all results is called the *sample space* of the experiment. The *power set* of the sample space is formed by considering all different collections of possible results. For example, rolling a die produces one of six possible results. One collection of possible results corresponds to getting an odd number.

Thus, the subset {1,3,5} is an element of the power set of the sample space of die rolls. These collections are called *events*. In this case, {1,3,5} is the event that the die falls on some odd number. If the results that actually occur fall in a given event, that event is said to have occurred.

Probability is a way of assigning every "event" a value between zero and one, with the requirement that the event made up of all possible results (in our example, the event {1,2,3,4,5,6}) be assigned a value of one. To qualify as a probability distribution, the assignment of values must satisfy the requirement that if you look at a collection of mutually exclusive events (events that contain no common results, e.g., the events {1,6}, {3}, and {2,4} are all mutually exclusive), the probability that at least one of the events will occur is given by the sum of the probabilities of all the individual events.

The probability that any one of the events {1,6}, {3}, or {2,4} will occur is 5/6. This is the same as saying that the probability of event {1,2,3,4,6} is 5/6. This event encompasses the possibility of any number except five being rolled. The mutually exclusive event {5} has a probability of 1/6, and the event {1,2,3,4,5,6} has a probability of 1, that is, absolute certainty.

Measure-theoretic Probability Theory

The *raison d'être* of the measure-theoretic treatment of probability is that it unifies the discrete and the continuous cases, and makes the difference a question of which measure is used. Furthermore, it covers distributions that are neither discrete nor continuous nor mixtures of the two.

An example of such distributions could be a mix of discrete and continuous distributions—for example, a random variable that is 0 with probability 1/2, and takes a random value from a normal distribution with probability 1/2. It can still be studied to some extent by considering it to have a pdf of $(\delta[x]+\varphi(x))/2$, where $\delta[x]$ is the Dirac delta function.

Other distributions may not even be a mix, for example, the Cantor distribution has no positive probability for any single point, neither does it have a density. The modern approach to probability theory solves these problems using measure theory to define the probability space:

Given any set Ω, (also called sample space) and a σ-algebra $\mathcal{F}$ on it, a measure P defined on $\mathcal{F}$ is called a probability measure if $P(\Omega)=1$.

If $\mathcal{F}$ is the Borel σ-algebra on the set of real numbers, then there is a unique probability measure on $\mathcal{F}$ for any cdf, and vice versa. The measure corresponding to a cdf is said to be induced by the cdf. This measure coincides with the pmf for discrete variables and pdf for continuous variables, making the measure-theoretic approach free of fallacies.

The *probability* of a set E in the σ-algebra $\mathcal{F}$ is defined as

$$P(E) = \int_{\omega \in E} \mu_F(d\omega)$$

where the integration is with respect to the measure μ_F induced by F.

Along with providing better understanding and unification of discrete and continuous probabilities, measure-theoretic treatment also allows us to work on probabilities outside $\mathbb{R}^n$, as in the theory of stochastic processes. For example to study Brownian motion, probability is defined on a space of functions.

Probability Distribution

In probability and statistics, a probability distribution assigns a probability to each of the possible outcomes of a random experiment, survey, or procedure of statistical inference. Examples are found in experiments whose sample space is non-numerical, where the distribution would be a categorical distribution; experiments whose sample space is encoded by discrete random variables, where the distribution is a probability mass function; and experiments with sample spaces encoded by continuous random variables, where the distribution is a probability density function. More complex experiments, such as those involving stochastic processes defined in continuous-time, may demand the use of more general probability measures.

In applied probability, a probability distribution can be specified in a number of different ways, often chosen for mathematical convenience:

- by supplying a valid probability mass function or probability density function
- by supplying a valid cumulative distribution function or survival function
- by supplying a valid hazard function

- by supplying a valid characteristic function
- by supplying a rule for constructing a new random variable from other random variables whose joint probability distribution is known.

A probability distribution can either be univariate or multivariate. A univariate distribution gives the probabilities of a single random variable taking on various alternative values; a multivariate distribution (a joint probability distribution) gives the probabilities of a random vector — a set of two or more random variables — taking on various combinations of values.

Important and commonly encountered univariate probability distributions include the binomial distribution, the hypergeometric distribution, and the normal distribution. The multivariate normal distribution is a commonly encountered multivariate distribution.

Introduction

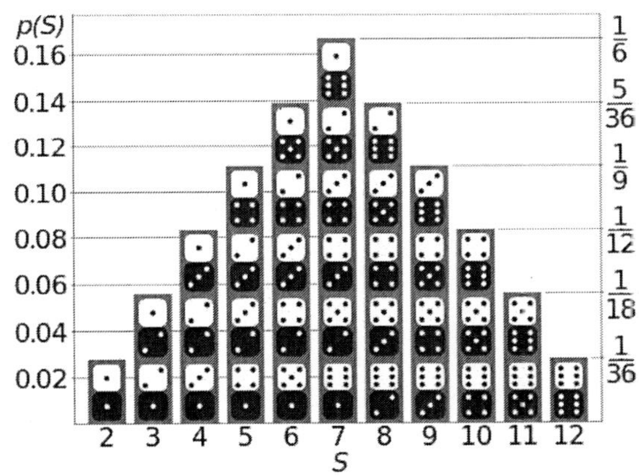

***Figure:** Discrete probability distribution for the sum of two dice.*

To define probability distributions for the simplest cases, one needs to distinguish between discrete and continuous random variables. In the discrete case, one can easily assign a probability to each possible value: for example, when throwing a die, each of the six values *1* to *6* has the probability 1/6. In contrast, when a random variable takes values from a continuum, probabilities are nonzero only if they refer to finite intervals: in quality control one might demand that the probability of a "500 g" package containing between 490 g and 510 g should be no less than 98%.

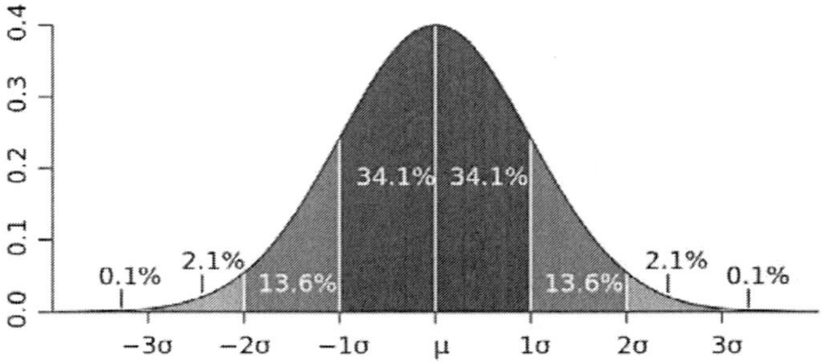

Figure: *Normal distribution, also called Gaussian or "bell curve", the most important continuous random distribution.*

If the random variable is real-valued (or more generally, if a total order is defined for its possible values), the cumulative distribution function gives the probability that the random variable is no larger than a given value; in the real-valued case it is the integral of the density.

Terminology

As probability theory is used in quite diverse applications, terminology is not uniform and sometimes confusing.

The following terms are used for non-cumulative probability distribution functions:

- Probability mass, Probability mass function, p.m.f.: for discrete random variables.
- Categorical distribution: for discrete random variables with a finite set of values.
- Probability density, Probability density function, p.d.f: most often reserved for continuous random variables.

The following terms are somewhat ambiguous as they can refer to non-cumulative or cumulative distributions, depending on authors' preferences:

- Probability distribution function: continuous or discrete, non-cumulative or cumulative.

- Probability function: even more ambiguous, can mean any of the above or other things.

Finally,

- Probability distribution: either the same as *probability distribution function*, or understood as something more fundamental underlying an actual mass or density function.

Basic Terms

- Mode: the most frequently occurring value in a discrete distribution, or the value in a continuous distribution at which its probability density function has its maximum value.
- Tail: a region of extreme values in a distribution.
- Support: the smallest closed set whose complement has probability zero; the set of points or elements that are actual members of the distribution.
- Expected value or mean: the weighted average of the possible values, using their probabilities as their weights; or, the continuous analog thereof.
- Median: the value which there is a 50% chance of being either below or above.
- Variance: an important measure of the dispersion of the distribution.
- Standard deviation: the square root of the variance, and hence another measure of dispersion.
- Symmetry: a property of some distributions in which the portion of the distribution to the left of a specific value is a mirror image of the portion to its right.
- Skewness: a measure of the extent to which a distribution "leans" to one side of its mean.

Cumulative Distribution Function

Because a probability distribution Pr on the real line is determined by the probability of a scalar random variable X being in a half-open interval $(-\infty, x]$, the probability distribution is completely characterized by its cumulative distribution function:

$$F(x) = \Pr\left[X \leq x\right] \qquad \text{for all } x \in \mathbb{R}.$$

Discrete Probability Distribution

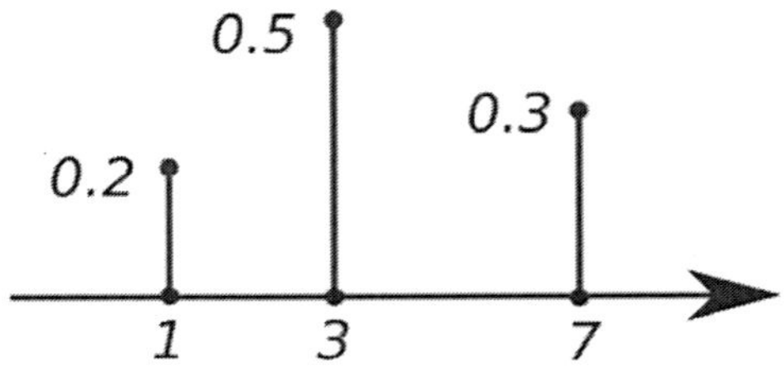

Figure: *The probability mass function of a discrete probability distribution. The probabilities of the singletons {1}, {3}, and {7} are respectively 0.2, 0.5, 0.3. A set not containing any of these points has probability zero.*

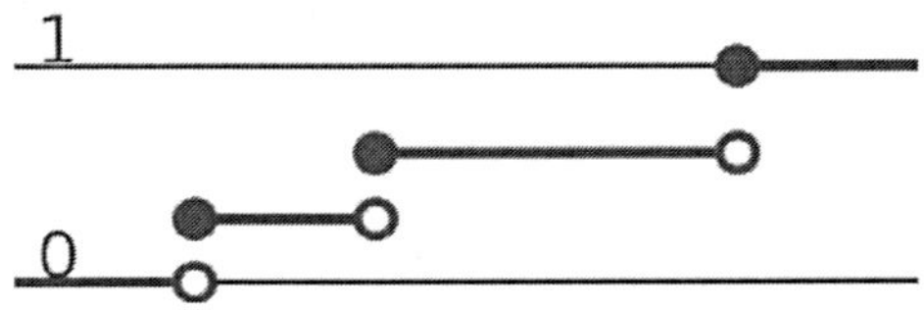

Figure: *The cdf of a discrete probability distribution, ...*

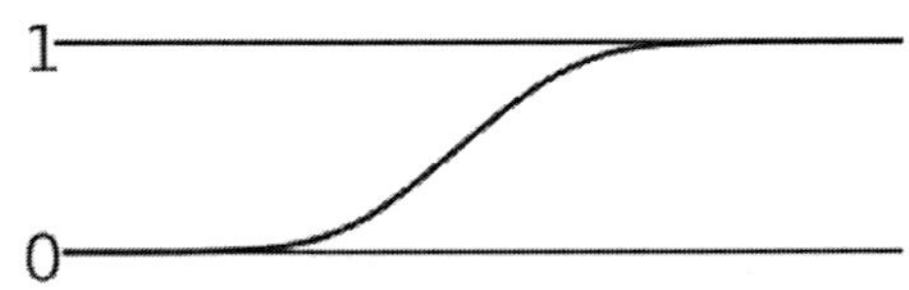

Figure: *... of a continuous probability distribution, ...*

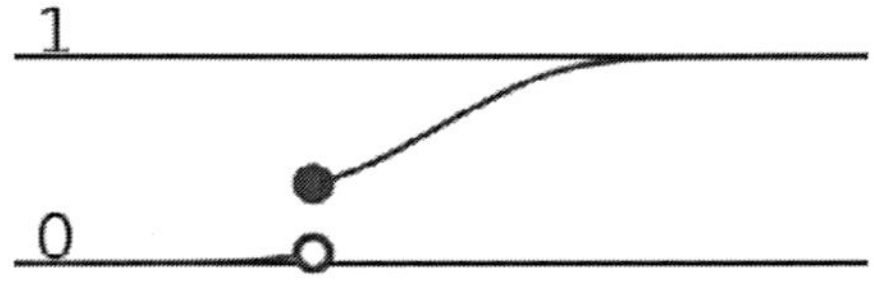

Figure: *... of a distribution which has both a continuous part and a discrete part.*

A discrete probability distribution shall be understood as a *probability distribution* characterized by a probability mass function. Thus, the distribution of a random variable X is discrete, and X is then called a discrete random variable, if

$$\sum_u \Pr(X = u) = 1$$

as u runs through the set of all possible values of X. It follows that such a random variable can assume only a finite or countably infinite number of values. For the number of potential values to be countably infinite even though their probabilities sum to 1 requires that the probabilities decline to zero fast enough: for example, if $\Pr(X = n) = \frac{1}{2^n}$ for $n = 1, 2, ...,$ we have the sum of probabilities 1/2 + 1/4 + 1/8 + ... = 1.

In cases more frequently considered, this set of possible values is a topologically discrete set in the sense that all its points are isolated points. But there are discrete random variables for which this countable set is dense on the real line (for example, a distribution over rational numbers).

Among the most well-known discrete probability distributions that are used for statistical modelling are the Poisson distribution, the Bernoulli distribution, the binomial distribution, the geometric distribution, and the negative binomial distribution. In addition, the discrete uniform distribution is commonly used in computer programmes that make equal-probability random selections between a number of choices.

Cumulative Density

Equivalently to the above, a discrete random variable can be defined as a random variable whose cumulative distribution function (cdf) increases only by jump discontinuities—that is, its cdf increases only where it "jumps" to a higher value, and is constant between those jumps. The points where jumps occur are precisely the values which the random variable may take.

Delta-function Representation

Consequently, a discrete probability distribution is often represented as a generalized probability density function involving Dirac delta functions, which substantially unifies the treatment of continuous and discrete distributions. This is especially useful when dealing with probability distributions involving both a continuous and a discrete part.

Indicator-function Representation

For a discrete random variable X, let u_0, u_1, ... be the values it can take with non-zero probability. Denote

$$\Omega_i = \{\omega : X(\omega) = u_i\}, i = 0,1,2,\ldots$$

These are disjoint sets, and by formula (1)

$$\Pr\left(\bigcup_i \Omega_i\right) = \sum_i \Pr(\Omega_i) = \sum_i \Pr(X = u_i) = 1.$$

It follows that the probability that X takes any value except for u_0, u_1, ... is zero, and thus one can write X as

$$X = \sum_i u_i 1_{\Omega_i}$$

except on a set of probability zero, where 1_A is the indicator function of A. This may serve as an alternative definition of discrete random variables.

Continuous Probability Distribution

A continuous probability distribution is a *probability distribution* that has a probability density function. Mathematicians also call such a distribution absolutely continuous, since its cumulative distribution function is absolutely continuous with respect to the Lebesgue measure λ. If the distribution of X is continuous, then X is called a continuous random variable. There are many examples of continuous probability distributions: normal, uniform, chi-squared, and others.

Intuitively, a continuous random variable is the one which can take a continuous range of values — as opposed to a discrete distribution, where the set of possible values for the random variable is at most countable. While for a discrete distribution an event with probability zero is impossible (e.g. rolling 3½ on a standard die is impossible, and has probability zero), this is not so in the case of a continuous random variable.

For example, if one measures the width of an oak leaf, the result of 3½ cm is possible, however it has probability zero because there are uncountably many other potential values even between 3 cm and 4 cm.

Each of these individual outcomes has probability zero, yet the probability that the outcome will fall into the interval (3 cm, 4 cm) is nonzero. This apparent paradox is resolved by the fact that the probability that X attains some value within an infinite set, such as an interval, cannot be found by naively adding the probabilities for individual values. Formally, each value has an infinitesimally small probability, which statistically is equivalent to zero.

Formally, if X is a continuous random variable, then it has a probability density function $f(x)$, and therefore its probability of falling into a given interval, say $[a, b]$ is given by the integral

$$\Pr[a \le X \le b] = \int_a^b f(x)dx$$

In particular, the probability for X to take any single value a (that is $a \le X \le a$) is zero, because an integral with coinciding upper and lower limits is always equal to zero.

The definition states that a continuous probability distribution must possess a density, or equivalently, its cumulative distribution function be absolutely continuous. This requirement is stronger than simple continuity of the cumulative distribution function, and there is a special class of distributions, singular distributions, which are neither continuous nor discrete nor a mixture of those. An example is given by the Cantor distribution. Such singular distributions however are never encountered in practice.

Note on terminology: some authors use the term "continuous distribution" to denote the distribution with continuous cumulative distribution function. Thus, their definition includes both the (absolutely) continuous and singular distributions.

By one convention, a probability distribution μ is called *continuous* if its cumulative distribution function $F(x) = \mu(-\infty, x]$ is continuous and, therefore, the probability measure of singletons $\mu\{x\}=0$ for all x.

Another convention reserves the term *continuous probability distribution* for absolutely continuous distributions. These distributions can be characterized by a probability density function: a non-negative Lebesgue integrable function f defined on the real numbers such that

$$F(x) = \mu(-\infty, x] = \int_{-\infty}^{x} f(t)dt.$$

Discrete distributions and some continuous distributions (like the Cantor distribution) do not admit such a density.

Some Properties

- The probability distribution of the sum of two independent random variables is the convolution of each of their distributions.
- Probability distributions are not a vector space – they are not closed under linear combinations, as these do not preserve non-negativity or total integral 1 – but they are closed under convex combination, thus forming a convex subset of the space of functions (or measures).

Kolmogorov definition

In the measure-theoretic formalization of probability theory, a random variable is defined as a measurable function X from a probability space $(\Omega, \mathcal{F}, \mathrm{P})$ to measurable space $(\mathcal{X}, \mathcal{A})$. A probability distribution is the pushforward measure, P, satisfying $X_*\mathrm{P} = \mathrm{P}X^{-1}$ on .

Random Number Generation

A frequent problem in statistical simulations (the Monte Carlo method) is the generation of pseudo-random numbers that are distributed in a given way. Most algorithms are based on a pseudorandom number generator that produces numbers X that are uniformly distributed in the interval [0,1). These random variates X are then transformed via some algorithm to create a new random variate having the required probability distribution.

Applications

The concept of the probability distribution and the random variables which they describe underlies the mathematical discipline of probability theory, and the science of statistics. There is spread or variability in almost any value that can be measured in a population (e.g. height of people, durability of a metal, sales growth, traffic flow, etc.); almost all measurements are made with some intrinsic error; in physics many processes are described probabilistically, from the kinetic properties of gases to the quantum mechanical description of fundamental particles. For these and many other reasons, simple numbers are often inadequate for describing a quantity, while probability distributions are often more appropriate.

As a more specific example of an application, the cache language models and other statistical language models used in natural language processing to assign probabilities to the occurrence of particular words and word sequences do so by means of probability distributions.

Common Probability Distributions

The following is a list of some of the most common probability distributions, grouped by the type of process that they are related to. Note also that all of the univariate distributions below are singly peaked; that is, it is assumed that the values cluster around a single point. In practice, actually observed quantities may cluster around multiple values. Such quantities can be modelled using a mixture distribution.

Related to real-valued quantities that grow linearly (e.g. errors, offsets)

- Normal distribution (Gaussian distribution), for a single such quantity; the most common continuous distribution

Related to positive real-valued quantities that grow exponentially (e.g. prices, incomes, populations)

- Log-normal distribution, for a single such quantity whose log is normally distributed
- Pareto distribution, for a single such quantity whose log is exponentially distributed; the prototypical power law distribution

Related to real-valued quantities that are assumed to be uniformly distributed over a (possibly unknown) region

- Discrete uniform distribution, for a finite set of values (e.g. the outcome of a fair die)
- Continuous uniform distribution, for continuously distributed values

Related to Bernoulli trials (yes/no events, with a given probability)

- Basic distributions:
  * Bernoulli distribution, for the outcome of a single Bernoulli trial (e.g. success/failure, yes/no)

  * Binomial distribution, for the number of "positive occurrences" (e.g. successes, yes votes, etc.) given a fixed total number of independent occurrences
  * Negative binomial distribution, for binomial-type observations but where the quantity of interest is the number of failures before a given number of successes occurs
  * Geometric distribution, for binomial-type observations but where the quantity of interest is the number of failures before the first success; a special case of the negative binomial distribution
- Related to sampling schemes over a finite population:
  * Hypergeometric distribution, for the number of "positive occurrences" (e.g. successes, yes votes, etc.) given a fixed number of total occurrences, using sampling without replacement
  * Beta-binomial distribution, for the number of "positive occurrences" (e.g. successes, yes votes, etc.) given a fixed number of total occurrences, sampling using a Polya urn scheme (in some sense, the "opposite" of sampling without replacement)

Related to categorical outcomes (events with K possible outcomes, with a given probability for each outcome)

- Categorical distribution, for a single categorical outcome (e.g. yes/no/maybe in a survey); a generalization of the Bernoulli distribution
- Multinomial distribution, for the number of each type of categorical outcome, given a fixed number of total outcomes; a generalization of the binomial distribution
- Multivariate hypergeometric distribution, similar to the multinomial distribution, but using sampling without replacement; a generalization of the hypergeometric distribution

Related to events in a Poisson process (events that occur independently with a given rate)

- Poisson distribution, for the number of occurrences of a Poisson-type event in a given period of time

- Exponential distribution, for the time before the next Poisson-type event occurs

Useful for hypothesis testing related to normally distributed outcomes

- Chi-squared distribution, the distribution of a sum of squared standard normal variables; useful e.g. for inference regarding the sample variance of normally distributed samples
- Student's t distribution, the distribution of the ratio of a standard normal variable and the square root of a scaled chi squared variable; useful for inference regarding the mean of normally distributed samples with unknown variance
- F-distribution, the distribution of the ratio of two scaled chi squared variables; useful e.g. for inferences that involve comparing variances or involving R-squared (the squared correlation coefficient)

Useful as conjugate prior distributions in Bayesian inference

- Beta distribution, for a single probability (real number between 0 and 1); conjugate to the Bernoulli distribution and binomial distribution
- Gamma distribution, for a non-negative scaling parameter; conjugate to the rate parameter of a Poisson distribution or exponential distribution, the precision (inverse variance) of a normal distribution, etc.
- Dirichlet distribution, for a vector of probabilities that must sum to 1; conjugate to the categorical distribution and multinomial distribution; generalization of the beta distribution
- Wishart distribution, for a symmetric non-negative definite matrix; conjugate to the inverse of the covariance matrix of a multivariate normal distribution; generalization of the gamma distribution

Convergence of Random Variables

In probability theory, there exist several different notions of convergence of random variables. The convergence of sequences of random variables to some limit random variable is an important concept in probability theory, and its applications to statistics and stochastic processes. The same concepts are known in more general mathematics as stochastic convergence and they formalize the idea that a sequence of essentially random or unpredictable events can sometimes be expected to settle down into a behaviour that is essentially unchanging when items far enough into the sequence are studied. The different possible notions of convergence relate to how such a behaviour can be characterised: two readily understood behaviours are that the sequence eventually takes a constant value, and that values in the sequence continue to change but can be described by an unchanging probability distribution.

Background

"Stochastic convergence" formalizes the idea that a sequence of essentially random or unpredictable events can sometimes be expected to settle into a pattern. The pattern may for instance be

- Convergence in the classical sense to a fixed value, perhaps itself coming from a random event
- An increasing similarity of outcomes to what a purely deterministic function would produce
- An increasing preference towards a certain outcome
- An increasing "aversion" against straying far away from a certain outcome

Some less obvious, more theoretical patterns could be

- That the probability distribution describing the next outcome may grow increasingly similar to a certain distribution
- That the series formed by calculating the expected value of the outcome's distance from a particular value may converge to 0
- That the variance of the random variable describing the next event grows smaller and smaller.

These other types of patterns that may arise are reflected in the different types of stochastic convergence that have been studied.

While the above discussion has related to the convergence of a single series to a limiting value, the notion of the convergence of two series towards each other is also important, but this is easily handled by studying the sequence defined as either the difference or the ratio of the two series.

For example, if the average of n uncorrelated random variables Y_i, $i = 1, \ldots, n$, all having the same finite mean and variance, is given by

$$X_n = \frac{1}{n}\sum_{i=1}^{n} Y_i,$$

then as n tends to infinity, X_n converges *in probability* to the common mean, μ, of the random variables Y_i. This result is known as the weak law of large numbers. Other forms of convergence are important in other useful theorems, including the central limit theorem.

Throughout the following, we assume that (X_n) is a sequence of random variables, and X is a random variable, and all of them are defined on the same probability space $(\Omega, \mathcal{F}, P)$.

Convergence in Distribution

Examples of Convergence in Distribution

Dice factory: Suppose a new dice factory has just been built. The first few dice come out quite biased, due to imperfections in the production process. The outcome from tossing any of them will follow a distribution markedly different from the desired uniform distribution.

As the factory is improved, the dice become less and less loaded, and the outcomes from tossing a newly produced dice will follow the uniform distribution more and more closely.

Tossing Coins

Let X_n be the fraction of heads after tossing up an unbiased coin n times. Then X_1 has the Bernoulli distribution with expected value $\mu = 0.5$ and variance $\sigma^2 = 0.25$. The subsequent random variables X_2, X_3, ... will all be distributed binomially.

As n grows larger, this distribution will gradually start to take shape more and more similar to the bell curve of the normal distribution. If we shift and rescale X_n's appropriately, then $Z_n = \sqrt{n}(X_n - \mu) / \sigma$ will be converging in distribution to the standard normal, the result that follows from the celebrated central limit theorem.

Graphic Example

Suppose $\{X_i\}$ is an iid sequence of uniform $U(“1,1)$ random variables. Let $Z_n = \frac{1}{\sqrt{n}}\sum_{i=1}^{n} X_i$ be their (normalized) sums. Then according to the central limit theorem, the distribution of Z_n approaches the normal $N(0,^1/_3)$ distribution. This convergence is shown in the picture: as n grows larger, the shape of the pdf function gets closer and closer to the Gaussian curve.

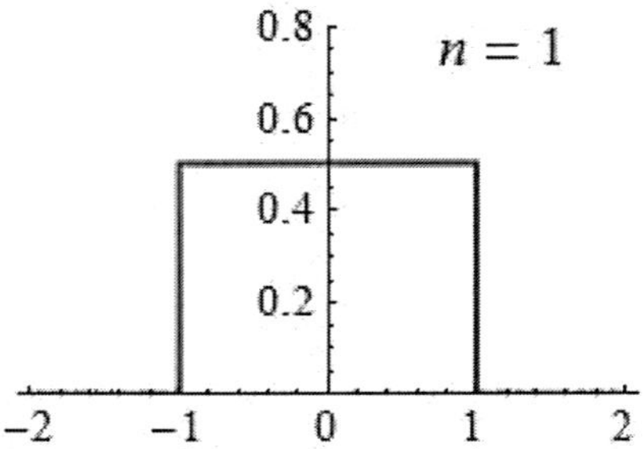

With this mode of convergence, we increasingly expect to see the next outcome in a sequence of random experiments becoming better and better modelled by a given probability distribution.

Convergence in distribution is the weakest form of convergence, since it is implied by all other types of convergence mentioned in this article. However convergence in distribution is very frequently used in practice; most often it arises from application of the central limit theorem.

Properties

- Since $F(a) = \Pr(X \le a)$, the convergence in distribution means that the probability for X_n to be in a given range is approximately equal to the probability that the value of X is in that range, provided n is sufficiently large.
- In general, convergence in distribution does not imply that the sequence of corresponding probability density functions will also converge. As an example one may

consider random variables with densities $f_n(x) = (1 - \cos(2\pi nx))1_{\{x-(0,1)\}}$. These random variables converge in distribution to a uniform $U(0, 1)$, whereas their densities do not converge at all. However, Scheffe's theorem states that convergence of the probability density functions implies convergence in distribution.

- The portmanteau lemma provides several equivalent definitions of convergence in distribution. Although these definitions are less intuitive, they are used to prove a number of statistical theorems. The lemma states that $\{X_n\}$ converges in distribution to X if and only if any of the following statements are true:

* $\mathrm{E}f(X_n) \to \mathrm{E}f(X)$ for all bounded, continuous functions f;
* $\mathrm{E}f(X_n) \to \mathrm{E}f(X)$ for all bounded, Lipschitz functions f;
* $\limsup\{ \mathrm{E}f(X_n) \} \leq \mathrm{E}f(X)$ for every upper semi-continuous function f bounded from above;
* $\liminf\{ \mathrm{E}f(X_n) \} \geq \mathrm{E}f(X)$ for every lower semi-continuous function f bounded from below;
* $\limsup\{ \Pr(X_n \in C) \} \leq \Pr(X \in C)$ for all closed sets C;
* $\liminf\{ \Pr(X_n \in U) \} \geq \Pr(X \in U)$ for all open sets U;
* $\lim\{ \Pr(X_n \in A) \} = \Pr(X \in A)$ for all continuity sets A of random variable X.

- The continuous mapping theorem states that for a continuous function $g(.)$, if the sequence $\{X_n\}$ converges in distribution to X, then so does $\{g(X_n)\}$ converge in distribution to $g(X)$.
- Lévy's continuity theorem: the sequence $\{X_n\}$ converges in distribution to X if and only if the sequence of corresponding characteristic functions $\{\varphi_n\}$ converges pointwise to the characteristic function φ of X.
- Convergence in distribution is metrizable by the Lévy–Prokhorov metric.
- A natural link to convergence in distribution is the Skorokhod's representation theorem.

Chapter 6

Convergence in Probability

Examples of Convergence in Probability

Height of a Person

Consider the following experiment. First, pick a random person in the street. Let X be his/her height, which is *ex ante* a random variable. Then you start asking other people to estimate this height by eye.

Let X_n be the average of the first n responses. Then (provided there is no systematic error) by the law of large numbers, the sequence X_n will converge in probability to the random variable X.

Archer

Suppose a person takes a bow and starts shooting arrows at a target. Let X_n be his score in n-th shot. Initially he will be very likely to score zeros, but as the time goes and his archery skill increases, he will become more and more likely to hit the bullseye and score 10 points.

After the years of practice the probability that he hit anything but 10 will be getting increasingly smaller and smaller. Thus, the sequence X_n converges in probability to $X = 10$.

Note that X_n does not converge almost surely however. No matter how professional the archer becomes, there will always be a small probability of making an error.

Thus the sequence $\{X_n\}$ will never turn stationary: there will always be non-perfect scores in it, even if they are becoming increasingly less frequent.

The basic idea behind this type of convergence is that the probability of an "unusual" outcome becomes smaller and smaller as the sequence progresses.

The concept of convergence in probability is used very often in statistics.

For example, an estimator is called consistent if it converges in probability to the quantity being estimated. Convergence in probability is also the type of convergence established by the weak law of large numbers.

Definition

A sequence $\{X_n\}$ of random variables converges in probability towards X if for all $\varepsilon > 0$

$$\lim_{n\to\infty} \Pr\left(|X_n - X| \geq \varepsilon\right) = 0.$$

Formally, pick any $\varepsilon > 0$ and any $\delta > 0$. Let P_n be the probability that X_n is outside the ball of radius ε centred at X.

Then for X_n to converge in probability to X there should exist a number N_δ such that for all $n \geq N_\delta$ the probability P_n is less than δ.

Convergence in probability is denoted by adding the letter p over an arrow indicating convergence, or using the "plim" probability limit operator:

$$X_n \xrightarrow{p} X,\ X_n \xrightarrow{P} X,\ \underset{n\to\infty}{\operatorname{plim}} X_n = X.$$

For random elements $\{X_n\}$ on a separable metric space (S, d), convergence in probability is defined similarly by

$$\forall \varepsilon > 0, \Pr\left(d(X_n, X) \geq \varepsilon\right) \to 0.$$

Properties

- Convergence in probability implies convergence in distribution.
- Convergence in probability does not imply almost sure convergence.
- In the opposite direction, convergence in distribution implies convergence in probability only when the limiting random variable X is a constant.

- The continuous mapping theorem states that for every continuous function $g(\cdot)$, if $X_n \xrightarrow{p} X$, then also $g(X_n) \xrightarrow{p} g(X)$.
- Convergence in probability defines a topology on the space of random variables over a fixed probability space. This topology is metrizable by the *Ky Fan metric*:

$$d(X,Y) = \inf\{\varepsilon > 0 : \Pr(|X - Y| > \varepsilon) \le \varepsilon\}$$

or

$$d(X,Y) = \mathbb{E}[\min(|X - Y|,1)].$$

Almost Sure Convergence

Examples of Almost Sure Convergence

Example 1

Consider an animal of some short-lived species. We record the amount of food that this animal consumes per day.

This sequence of numbers will be unpredictable, but we may be *quite certain* that one day the number will become zero, and will stay zero forever after.

Example 2

Consider a man who tosses seven coins every morning. Each afternoon, he donates one pound to a charity for each head that appeared.

The first time the result is all tails, however, he will stop permanently.

Let X_1, X_2, ... be the daily amounts the charity receives from him.

We may be *almost sure* that one day this amount will be zero, and stay zero forever after that.

However, when we consider *any finite number* of days, there is a nonzero probability the terminating condition will not occur.

This is the type of stochastic convergence that is most similar to pointwise convergence known from elementary real analysis.

Definition

To say that the sequence X_n converges almost surely or almost everywhere or with probability 1 or strongly towards X means that

$$\Pr\left(\lim_{n\to\infty} X_n = X\right) = 1.$$

This means that the values of X_n approach the value of X, in the sense that events for which X_n does not converge to X have probability 0. Using the probability space $(\Omega, \mathcal{F}, P)$ and the concept of the random variable as a function from Ω to R, this is equivalent to the statement

$$\Pr\left(\omega \in \Omega : \lim_{n\to\infty} X_n(\omega) = X(\omega)\right) = 1.$$

Another, equivalent, way of defining almost sure convergence is as follows:

$$\lim_{n\to\infty} \Pr\left(\omega \in \Omega : \sup_{m\geq n} |X_m(\omega) - X(\omega)| \geq \varepsilon\right) = 0 \quad \text{for all} \quad \varepsilon > 0.$$

Almost sure convergence is often denoted by adding the letters *a.s.* over an arrow indicating convergence:

$$X_n \xrightarrow{\text{a.s.}} X.$$

For generic random elements $\{X_n\}$ on a metric space (S, d), convergence almost surely is defined similarly:

$$\Pr\left(\omega \in \Omega : d\left(X_n(\omega), X(\omega)\right) \underset{n\to\infty}{\longrightarrow} 0\right) = 1$$

Properties

- Almost sure convergence implies convergence in probability, and hence implies convergence in distribution. It is the notion of convergence used in the strong law of large numbers.
- The concept of almost sure convergence does not come from a topology on the space of random variables. This means there is no topology on the space of random variables such that the almost surely convergent sequences are exactly the converging sequences with respect to that topology. In particular, there is no metric of almost sure convergence.

Sure Convergence

To say that the sequence or random variables (X_n) defined over the same probability space (i.e., a random process) converges surely or everywhere or pointwise towards X means

$$\lim_{n\to\infty} X_n(\omega) = X(\omega), \forall\omega \in \Omega.$$

where Ω is the sample space of the underlying probability space over which the random variables are defined.

This is the notion of pointwise convergence of sequence functions extended to sequence of random variables. (Note that random variables themselves are functions).

$$\left\{\omega \in \Omega | \lim_{n\to\infty} X_n(\omega) = X(\omega)\right\} = \Omega.$$

Sure convergence of a random variable implies all the other kinds of convergence stated above, but there is no payoff in probability theory by using sure convergence compared to using almost sure convergence. The difference between the two only exists on sets with probability zero. This is why the concept of sure convergence of random variables is very rarely used.

Convergence in Mean

We say that the sequence X_n converges in the r-th mean (or in the L^r-norm) towards X, for some $r \geq 1$, if r-th absolute moments of X_n and X exist, and

$$\lim_{n\to\infty} \mathrm{E}\left(| X_n - X |^r\right) = 0,$$

where the operator E denotes the expected value. Convergence in r-th mean tells us that the expectation of the r-th power of the difference between X_n and X converges to zero.

This type of convergence is often denoted by adding the letter L^r over an arrow indicating convergence:

$$X_n \xrightarrow{L^r} X.$$

The most important cases of convergence in r-th mean are:

- When X_n converges in r-th mean to X for $r = 1$, we say that X_n converges in mean to X.
- When X_n converges in r-th mean to X for $r = 2$, we say that X_n converges in mean square to X. This is also

sometimes referred to as convergence in mean, and is sometimes denoted

$$\underset{n\to\infty}{\text{l.i.m.}} X_n = X.$$

Convergence in the r-th mean, for $r \geq 1$, implies convergence in probability (by Markov's inequality), while if $r > s \geq 1$, convergence in r-th mean implies convergence in s-th mean. Hence, convergence in mean square implies convergence in mean.

Properties

The chain of implications between the various notions of convergence are noted in their respective sections. They are, using the arrow notation:

$$\begin{array}{ccccc} \xrightarrow{L^s} & \underset{s>r\geq 1}{\Rightarrow} & \xrightarrow{L^r} & & \\ & & \Downarrow & & \\ \xrightarrow{a.s.} & \Rightarrow & \xrightarrow{p} & \Rightarrow & \xrightarrow{d} \end{array}$$

These properties, together with a number of other special cases, are summarized in the following list:

- Almost sure convergence implies convergence in probability:

$$X_n \xrightarrow{as} X \;\Rightarrow\; X_n \xrightarrow{p} X$$

- Convergence in probability implies there exists a sub-sequence (k_n) which almost surely converges:

$$X_n \xrightarrow{p} X \;\Rightarrow\; X_{k_n} \xrightarrow{as} X$$

- Convergence in probability implies convergence in distribution:

$$X_n \xrightarrow{p} X \;\Rightarrow\; X_n \xrightarrow{d} X$$

- Convergence in r-th order mean implies convergence in probability:

$$X_n \xrightarrow{L^r} X \;\Rightarrow\; X_n \xrightarrow{p} X$$

- Convergence in r-th order mean implies convergence in lower order mean, assuming that both orders are greater or equal than one:

$$X_n \xrightarrow{L^r} X \;\Rightarrow\; X_n \xrightarrow{L^s} X, \text{ provided } r \geq s \geq 1.$$

- If X_n converges in distribution to a constant c, then X_n converges in probability to c:

$$X_n \xrightarrow{d} c \;\Rightarrow\; X_n \xrightarrow{p} c, \text{ provided } c \text{ is a constant.}$$

If X_n converges in distribution to X and the difference between X_n and Y_n converges in probability to zero, then Y_n also converges in distribution to

$$X\!: X_n \xrightarrow{d} X,\ |X_n - Y_n| \xrightarrow{p} 0 \;\Rightarrow\; Y_n \xrightarrow{d} X$$

- If X_n converges in distribution to X and Y_n converges in distribution to a constant c, then the joint vector (X_n, Y_n) converges in distribution to (X, c):

$$X_n \xrightarrow{d} X,\ Y_n \xrightarrow{d} c \;\Rightarrow\; (X_n, Y_n) \xrightarrow{d} (X, c)$$

  provided c is a constant.

  Note that the condition that Y_n converges to a constant is important, if it were to converge to a random variable Y then we wouldn't be able to conclude that (X_n, Y_n) converges to (X, Y).

- If X_n converges in probability to X and Y_n converges in probability to Y, then the joint vector (X_n, Y_n) converges in probability to (X, Y):

$$X_n \xrightarrow{p} X,\ Y_n \xrightarrow{p} Y \;\Rightarrow\; (X_n, Y_n) \xrightarrow{p} (X, Y)$$

- If X_n converges in probability to X, and if $P(|X_n| \leq b) = 1$ for all n and some b, then X_n converges in rth mean to X for all $r \geq 1$. In other words, if X_n converges in probability to X and all random variables X_n are almost surely bounded above and below, then X_n converges to X also in any rth mean.

- Almost sure representation. Usually, convergence in distribution does not imply convergence almost surely. However for a given sequence $\{X_n\}$ which converges in distribution to X_0 it is always possible to find a new probability space (Ω, F, P) and random variables $\{Y_n, n = 0, 1, \ldots\}$ defined on it such that Y_n is equal in distribution to X_n for each $n \geq 0$, and Y_n converges to Y_0 almost surely.

- If for all $\varepsilon > 0$,

$$\sum_n \mathbb{P}(|X_n - X| > \varepsilon) < \infty,$$

then we say that X_n *converges almost completely*, or *almost in probability* towards X. When X_n converges almost completely towards X then it also converges almost surely to X. In other words, if X_n converges in probability to X sufficiently quickly (*i.e.* the above sequence of tail probabilities is summable for all $\varepsilon > 0$), then X_n also converges almost surely to X. This is a direct implication from the Borel-Cantelli lemma.

- If S_n is a sum of n real independent random variables:

$S_n = X_1 + \cdots + X_n$

then S_n converges almost surely if and only if S_n converges in probability.

- The dominated convergence theorem gives sufficient conditions for almost sure convergence to imply L^1-convergence:

$$\left.\begin{array}{l} X_n \xrightarrow{a.s.} X \\ |X_n| < Y \\ \mathrm{E}(Y) < \infty \end{array}\right\} \Rightarrow X_n \xrightarrow{L^1} X$$

- A necessary and sufficient condition for L^1 convergence is $X_n \xrightarrow{P} X$ and the sequence (X_n) is uniformly integrable.

Law of Large Numbers

In probability theory, the law of large numbers (LLN) is a theorem that describes the result of performing the same experiment a large number of times.

According to the law, the average of the results obtained from a large number of trials should be close to the expected value, and will tend to become closer as more trials are performed.

The LLN is important because it "guarantees" stable long-term results for the averages of random events. For example, while a casino may lose money in a single spin of the roulette wheel, its earnings will tend towards a predictable percentage over a large number of spins. Any winning streak by a player will eventually be overcome by the parameters of the game. It is important to remember that the LLN only applies (as the name indicates) when a *large number* of observations are considered. There is no principle that a small number of observations will coincide with the expected value or that a streak of one value will immediately be "balanced" by the others.

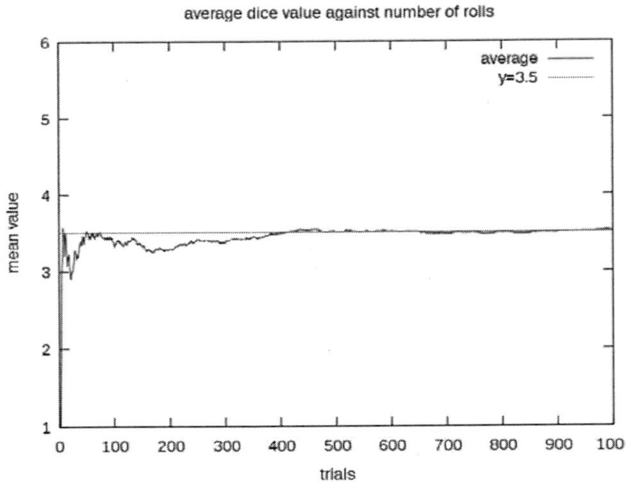

Figure: *An illustration of the law of large numbers using a particular run of rolls of a single die. As the number of rolls in this run increases, the average of the values of all the results approaches 3.5. While different runs would show a different shape over a small number of throws (at the left), over a large number of rolls (to the right) they would be extremely similar.*

Examples

For example, a single roll of a six-sided die produces one of the numbers 1, 2, 3, 4, 5, or 6, each with equal probability. Therefore, the expected value of a single die roll is

$$\frac{1+2+3+4+5+6}{6} = 3.5.$$

According to the law of large numbers, if a large number of six-sided dice are rolled, the average of their values (sometimes called the sample mean) is likely to be close to 3.5, with the accuracy increasing as more dice are rolled.

It follows from the law of large numbers that the empirical probability of success in a series of Bernoulli trials will converge to the theoretical probability. For a Bernoulli random variable, the expected value is the theoretical probability of success, and the average of n such variables (assuming they are independent and identically distributed (i.i.d.)) is precisely the relative frequency.

For example, a fair coin toss is a Bernoulli trial. When a fair coin is flipped once, the theoretical probability that the outcome will be heads is equal to 1/2. Therefore, according to the law of large numbers, the proportion of heads in a "large" number of coin flips "should be" roughly 1/2. In particular, the proportion of heads after n flips will almost surely converge to 1/2 as n approaches infinity.

Though the proportion of heads (and tails) approaches 1/ 2, almost surely the absolute (nominal) difference in the number of heads and tails will become large as the number of flips becomes large. That is, the probability that the absolute difference is a small number, approaches zero as the number of flips becomes large. Also, almost surely the ratio of the absolute difference to the number of flips will approach zero. Intuitively, expected absolute difference grows, but at a slower rate than the number of flips, as the number of flips grows.

History

The Italian mathematician Gerolamo Cardano (1501–1576) stated without proof that the accuracies of empirical statistics tend to improve with the number of trials. This was then formalized as a law of large numbers.

A special form of the LLN (for a binary random variable) was first proved by Jacob Bernoulli. It took him over 20 years to develop a sufficiently rigorous mathematical proof which was published in his *Ars Conjectandi* (The Art of Conjecturing) in 1713. He named this his "Golden Theorem" but it became generally known as "Bernoulli's Theorem". This should not be confused with the principle in physics with the same name, named after Jacob Bernoulli's nephew Daniel Bernoulli. In 1837, S.D. Poisson further described it under the name "la loi des grands nombres" ("The law of large numbers"). Thereafter, it was known under both names, but the "Law of large numbers" is most frequently used.

After Bernoulli and Poisson published their efforts, other mathematicians also contributed to refinement of the law, including Chebyshev, Markov, Borel, Cantelli and Kolmogorov and Khinchin, who finally provided a complete proof of the LLN for arbitrary random variables. These further studies have given rise to two prominent forms of the LLN. One is called the "weak" law and the other the "strong" law. These forms do not describe different laws but instead refer to different ways of describing the mode of convergence of the cumulative sample means to the expected value, and the strong form implies the weak.

Forms

Two different versions of the Law of Large Numbers are described below; they are called the *Strong Law of Large Numbers*, and the *Weak Law of Large Numbers*. Both versions of the law state that – with virtual certainty – the sample average

$$\bar{X}_n = \frac{1}{n}(X_1 + \cdots + X_n)$$

converges to the expected value

$$\bar{X}_n \to \mu \qquad \text{for} \qquad n \to \infty$$

where X_1, X_2, ... is an infinite sequence of i.i.d. integrable random variables with expected value $E(X_1) = E(X_2) = ... = \mu$. Integrability means that $E(|X_j|) < \infty$ for $j=1,2,...$.

An assumption of finite variance $\text{Var}(X_1) = \text{Var}(X_2) = ... = \sigma^2 < \infty$ is not necessary. Large or infinite variance will make the convergence slower, but the LLN holds anyway. This assumption is often used because it makes the proofs easier and shorter.

The difference between the strong and the weak version is concerned with the mode of convergence being asserted.

Weak law

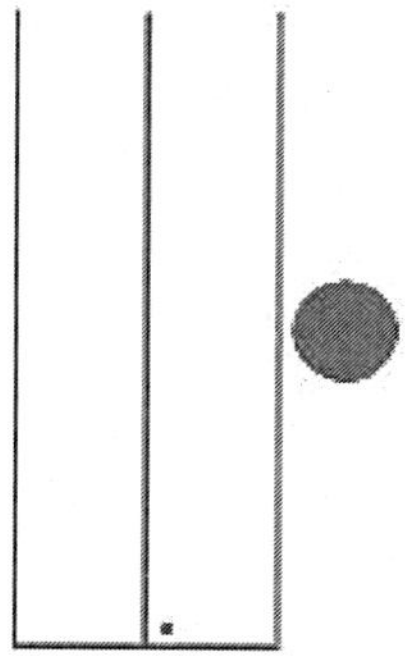

Figure: *Simulation illustrating the Law of Large Numbers. Each frame, you flip a coin that is red on one side and blue on the other, and put a dot in the corresponding column. A pie chart shows the proportion of red and blue so far. Notice that the proportion varies a lot at first, but gradually approaches 50%.*

The weak law of large numbers states that the sample average converges in probability towards the expected value

$$\bar{X}_n \xrightarrow{P} \mu \qquad \text{when } n \to \infty.$$

That is to say that for any positive number ε,

$$\lim_{n\to\infty} \Pr\left(|\bar{X}_n - \mu| > \varepsilon\right) = 0.$$

Interpreting this result, the weak law essentially states that for any nonzero margin specified, no matter how small, with a sufficiently large sample there will be a very high probability that the average of the observations will be close to the expected value, that is, within the margin.

Convergence in probability is also called weak convergence of random variables. This version is called the weak law because random variables may converge weakly (in probability) as above without converging strongly (almost surely) as below.

Strong Law

The strong law of large numbers states that the sample average converges almost surely to the expected value

$$\bar{X}_n \xrightarrow{a.s.} \mu \qquad \text{when } n \to \infty.$$

That is,

$$\Pr\left(\lim_{n\to\infty} \bar{X}_n = \mu\right) = 1.$$

The proof is more complex than that of the weak law. This law justifies the intuitive interpretation of the expected value of a random variable as the "long-term average when sampling repeatedly".

Almost sure convergence is also called strong convergence of random variables. This version is called the strong law because random variables which converge strongly (almost surely) are guaranteed to converge weakly (in probability). The strong law implies the weak law.

The strong law of large numbers can itself be seen as a special case of the pointwise ergodic theorem.

Moreover, if the summands are independent but not identically distributed, then

$$\bar{X}_n - \mathrm{E}\left[\bar{X}_n\right] \xrightarrow{a.s.} 0$$

provided that each X_k has a finite second moment and

$$\sum_{k=1}^{\infty} \frac{1}{k^2} \mathrm{Var}[X_k] < \infty.$$

This statement is known as *Kolmogorov's strong law.*

Differences between the Weak Law and the Strong Law

The *weak law* states that for a specified large n, the average $\bar{X}_n$ is likely to be near μ. Thus, it leaves open the possibility that $|\bar{X}_n - \mu| > \varepsilon$ happens an infinite number of times, although at infrequent intervals.

The *strong law* shows that this almost surely will not occur. In particular, it implies that with probability 1, we have that for any $\varepsilon > 0$ the inequality $|\bar{X}_n - \mu| < \varepsilon$ holds for all large enough n.

Uniform Law of Large Numbers

Suppose $f(x,\theta)$ is some function defined for $\theta \in \Theta$, and continuous in θ. Then for any fixed θ, the sequence $\{f(X_1,\theta), f(X_2,\theta), \ldots\}$ will be a sequence of independent and identically distributed random variables, such that the sample mean of this sequence converges in probability to E[$f(X,\theta)$]. This is the *pointwise* (in θ) convergence.

The uniform law of large numbers states the conditions under which the convergence happens *uniformly* in θ. If

1. Q is compact,
2. $f(x,\theta)$ is continuous at each $\theta \in \Theta$ for almost all x's, and measurable function of x at each θ.
3. there exists a dominating function $d(x)$ such that E[$d(X)$] $< \infty$, and

$$\|f(x,\theta)\| \leq d(x) \quad \text{for all } \theta \in \Theta.$$

Then E[$f(X,\theta)$] is continuous in θ, and

$$\sup_{\theta \in \Theta} \left\| \frac{1}{n} \sum_{i=1}^{n} f(X_i,\theta) - \mathrm{E}[f(X,\theta)] \right\| \xrightarrow{\text{a.s.}} 0.$$

Borel's Law of Large Numbers

Borel's law of large numbers, named after Émile Borel, states that if an experiment is repeated a large number of times, independently under identical conditions, then the proportion of times that any specified event occurs approximately equals the probability of the event's occurrence on any particular trial; the larger the number of repetitions, the better the approximation tends to be.

More precisely, if E denotes the event in question, p its probability of occurrence, and $N_n(E)$ the number of times E occurs in the first n trials, then with probability one,

$$\frac{N_n(E)}{n} \to p \text{ as } n \to \infty.$$

Chebyshev's Lemma. Let X be a random variable with finite expected value μ and finite non-zero variance σ^2. Then for any real number $k > 0$,

$$\Pr(|X - \mu| \geq k\sigma) \leq \frac{1}{k^2}.$$

This theorem makes rigorous the intuitive notion of probability as the long-run relative frequency of an event's occurrence. It is a special case of any of several more general laws of large numbers in probability theory.

Proof

Given $X_1, X_2, \ldots$ an infinite sequence of i.i.d. random variables with finite expected value $E(X_1) = E(X_2) = \ldots = \mu < \infty$, we are interested in the convergence of the sample average

$$\bar{X}_n = \tfrac{1}{n}(X_1 + \cdots + X_n).$$

The weak law of large numbers states:

Theorem: $\bar{X}_n \xrightarrow{P} \mu \quad \text{for} \quad n \to \infty.$

Proof Using Chebyshev's Inequality

This proof uses the assumption of finite variance $\mathrm{Var}(X_i) = \sigma^2$ (for all i). The independence of the random variables implies no correlation between them, and we have that

$$\mathrm{Var}(\bar{X}_n) = \mathrm{Var}(\tfrac{1}{n}(X_1 + \cdots + X_n)) = \frac{1}{n^2}\mathrm{Var}(X_1 + \cdots + X_n) = \frac{n\sigma^2}{n^2} = \frac{\sigma^2}{n}.$$

The common mean μ of the sequence is the mean of the sample average:

$$E(\bar{X}_n) = \mu.$$

Using Chebyshev's inequality on $\bar{X}_n$ results in

$$\mathrm{P}(\left|\bar{X}_n - \mu\right| \geq \varepsilon) \leq \frac{\sigma^2}{n\varepsilon^2}.$$

This may be used to obtain the following:

$$\mathrm{P}(\left|\bar{X}_n - \mu\right| < \varepsilon) = 1 - \mathrm{P}(\left|\bar{X}_n - \mu\right| \geq \varepsilon) \geq 1 - \frac{\sigma^2}{n\varepsilon^2}.$$

As n approaches infinity, the expression approaches 1. And by definition of convergence in probability, we have obtained

$$\bar{X}_n \xrightarrow{P} \mu \quad \text{for} \quad n \to \infty.$$

Proof Using Convergence of Characteristic Functions

By Taylor's theorem for complex functions, the characteristic function of any random variable, X, with finite mean μ, can be written as

$$\varphi_X(t) = 1 + it\mu + o(t), \quad t \to 0.$$

All X_1, X_2, ... have the same characteristic function, so we will simply denote this φ_X.

Among the basic properties of characteristic functions there are

$$\varphi_{\frac{1}{n}X}(t) = \varphi_X(\tfrac{t}{n}) \quad \text{and} \quad \varphi_{X+Y}(t) = \varphi_X(t)\varphi_Y(t)$$

if X and Y are independent.

These rules can be used to calculate the characteristic function of $\overline{X}_n$ in terms of φ_X:

$$\varphi_{\overline{X}_n}(t) = \left[\varphi_X\left(\frac{t}{n}\right)\right]^n = \left[1 + i\mu\frac{t}{n} + o\left(\frac{t}{n}\right)\right]^n \to e^{it\mu}, \quad \text{as} \quad n \to \infty.$$

The limit $e^{it\mu}$ is the characteristic function of the constant random variable μ, and hence by the Lévy continuity theorem, $\overline{X}_n$ converges in distribution to μ:

$$\overline{X}_n \xrightarrow{\mathcal{D}} \mu \qquad \text{for} \qquad n \to \infty.$$

μ is a constant, which implies that convergence in distribution to μ and convergence in probability to μ are equivalent. Therefore,

$$\overline{X}_n \xrightarrow{P} \mu \qquad \text{for} \qquad n \to \infty.$$

This shows that the sample mean converges in probability to the derivative of the characteristic function at the origin, as long as the latter exists.

Central Limit Theorem

In probability theory, the central limit theorem (CLT) states that, given certain conditions, the mean of a sufficiently large number of independent random variables, each with a well-defined mean and well-defined variance, will be approximately normally distributed. The central limit theorem has a number of variants.

In its common form, the random variables must be identically distributed. In variants, convergence of the mean to the normal distribution also occurs for non-identical distributions, given that they comply with certain conditions.

In more general probability theory, a central limit theorem is any of a set of weak-convergence theories. They all express the fact that a sum of many independent and identically distributed (i.i.d.) random variables, or alternatively, random variables with specific types of dependence, will tend to be distributed according to one of a small set of *attractor distributions*. When the variance of the i.i.d. variables is finite, the attractor distribution is the normal distribution. In contrast, the sum of a number of i.i.d. random variables with power law tail distributions decreasing as $|x|^{-\alpha-1}$ where $0 < \alpha < 2$ (and therefore having infinite variance) will tend to an alpha-stable distribution with stability parameter (or index of stability) of α as the number of variables grows.

Classical CLT

Let $\{X_1, ..., X_n\}$ be a random sample of size n—that is, a sequence of independent and identically distributed random variables drawn from distributions of expected values given by μ and finite variances given by σ^2. Suppose we are interested in the sample average

$$S_n := \frac{X_1 + \cdots + X_n}{n}$$

of these random variables. By the law of large numbers, the sample averages converge in probability and almost surely to the expected value μ as $n \to \infty$. The classical central limit theorem describes the size and the distributional form of the stochastic fluctuations around the deterministic number μ during this convergence.

More precisely, it states that as n gets larger, the distribution of the difference between the sample average S_n and its limit μ, when multiplied by the factor $\sqrt{n}$ (that is $\sqrt{n}(S_n - \mu)$), approximates the normal distribution with mean 0 and variance σ^2.

For large enough n, the distribution of S_n is close to the normal distribution with mean μ and variance $\sigma^2 n$. The usefulness of the theorem is that the distribution of $\sqrt{n}(S_n - \mu)$ approaches normality regardless of the shape of the distribution of the individual X_i's. Formally, the theorem can be stated as follows:

Lindeberg–Lévy CLT. Suppose $\{X_1, X_2, ...\}$ is a sequence of i.i.d. random variables with $E[X_i] = \mu$ and $Var[X_i] = \sigma^2 < \infty$. Then as n approaches infinity, the random variables $\sqrt{n}(S_n - \mu)$ converge in distribution to a normal $N(0, \sigma^2)$:

$$\sqrt{n}\left(\left(\frac{1}{n}\sum_{i=1}^{n} X_i\right) - \mu\right) \xrightarrow{d} N(0, \sigma^2).$$

In the case $\sigma > 0$, convergence in distribution means that the cumulative distribution functions of $\sqrt{n}(S_n - \mu)$ converge pointwise to the cdf of the $N(0, \sigma^2)$ distribution: for every real number z,

$$\lim_{n\to\infty} \Pr[\sqrt{n}(S_n - \mu) \le z] = \Phi(z / \sigma),$$

where $\Phi(x)$ is the standard normal cdf evaluated at x. Note that the convergence is uniform in z in the sense that

$$\lim_{n\to\infty} \sup_{z\in R} |\Pr[\sqrt{n}(S_n - \mu) \le z] - \Phi(z / \sigma)| = 0,$$

where sup denotes the least upper bound (or supremum) of the set.

Lyapunov CLT

The theorem is named after Russian mathematician Aleksandr Lyapunov. In this variant of the central limit theorem the random variables X_i have to be independent, but not necessarily identically distributed. The theorem also requires that random variables $|X_i|$ have moments of some order $(2 + \delta)$, and that the rate of growth of these moments is limited by the Lyapunov condition given below.

Lyapunov CLT. Suppose $\{X_1, X_2, ...\}$ is a sequence of independent random variables, each with finite expected value μ_i and variance $\sigma 2$

i . Define

$$s_n^2 = \sum_{i=1}^{n} \sigma_i^2$$

If for some $\delta > 0$, the *Lyapunov's condition*

$$\lim_{n\to\infty} \frac{1}{s_n^{2+\delta}} \sum_{i=1}^{n} E\left[|X_i - \mu_i|^{2+\delta}\right] = 0$$

is satisfied, then a sum of $(X_i - \mu_i)/s_n$ converges in distribution to a standard normal random variable, as n goes to infinity:

$$\frac{1}{s_n}\sum_{i=1}^{n}(X_i - \mu_i) \xrightarrow{d} \mathcal{N}(0,1).$$

In practice it is usually easiest to check the Lyapunov's condition for $\delta = 1$. If a sequence of random variables satisfies Lyapunov's condition, then it also satisfies Lindeberg's condition. The converse implication, however, does not hold.

Lindeberg CLT

In the same setting and with the same notation as above, the Lyapunov condition can be replaced with the following weaker one (from Lindeberg in 1920). For every $\varepsilon > 0$

$$\lim_{n\to\infty}\frac{1}{s_n^2}\sum_{i=1}^{n}\mathrm{E}\left[(X_i - \mu_i)^2 \cdot 1_{\{|X_i - \mu_i| > \varepsilon s_n\}}\right] = 0$$

where $1_{\{\ldots\}}$ is the indicator function. Then the distribution of the standardized sums $\frac{1}{s_n}\sum_{i=1}^{n}(X_i - \mu_i)$ converges towards the standard normal distribution N(0,1).

Multidimensional CLT

Proofs that use characteristic functions can be extended to cases where each individual $X_1, \ldots, X_n$ is an independent and identically distributed random vector in R^k, with mean vector $\mu = \mathrm{E}(X_i)$ and covariance matrix Σ (amongst the individual components of the vector). Now, if we take the summations of these vectors as being done componentwise, then the multidimensional central limit theorem states that when scaled, these converge to a multivariate normal distribution.

Let

$$\mathbf{X}_i = \begin{bmatrix} X_{i(1)} \\ \vdots \\ X_{i(k)} \end{bmatrix}$$

be the i-vector. The bold in $\mathbf{X}_i$ means that it is a random vector, not a random (univariate) variable. Then the sum of the random vectors will be

$$\begin{bmatrix} X_{1(1)} \\ \vdots \\ X_{1(k)} \end{bmatrix} + \begin{bmatrix} X_{2(1)} \\ \vdots \\ X_{2(k)} \end{bmatrix} + \cdots + \begin{bmatrix} X_{n(1)} \\ \vdots \\ X_{n(k)} \end{bmatrix} = \begin{bmatrix} \sum_{i=1}^{n} \left[X_{i(1)} \right] \\ \vdots \\ \sum_{i=1}^{n} \left[X_{i(k)} \right] \end{bmatrix} = \sum_{i=1}^{n} [\mathrm{X}_{\mathrm{i}}]$$

and the average will be

$$\left(\frac{1}{n}\right) \sum_{i=1}^{n} [\mathrm{X}_{\mathrm{i}}] = \frac{1}{n} \begin{bmatrix} \sum_{i=1}^{n} \left[X_{i(1)} \right] \\ \vdots \\ \sum_{i=1}^{n} \left[X_{i(k)} \right] \end{bmatrix} = \begin{bmatrix} \bar{X}_{i(1)} \\ \vdots \\ \bar{X}_{i(k)} \end{bmatrix} = \bar{\mathrm{X}}_{\mathrm{n}}$$

and therefore

$$\frac{1}{\sqrt{n}} \sum_{i=1}^{n} \left[\mathrm{X}_{\mathrm{i}} - E(X_i) \right] = \frac{1}{\sqrt{n}} \sum_{i=1}^{n} [\mathrm{X}_{\mathrm{i}} - \mu] = \sqrt{n} \left(\bar{\mathrm{X}}_n - \mu \right).$$

The multivariate central limit theorem states that

$$\sqrt{n} \left(\bar{\mathrm{X}}_n - \mu \right) \xrightarrow{D} \mathcal{N}_k(0, \Sigma)$$

where the covariance matrix Σ is equal to

$$\Sigma = \begin{bmatrix} Var(X_{1(1)}) & Cov(X_{1(1)}, X_{1(2)}) & Cov(X_{1(1)}, X_{1(3)}) & \cdots & Cov(X_{1(1)}, X_{1(k)}) \\ Cov(X_{1(2)}, X_{1(1)}) & Var(X_{1(2)}) & Cov(X_{1(2)}, X_{1(3)}) & \cdots & Cov(X_{1(2)}, X_{1(k)}) \\ Cov(X_{1(3)}, X_{1(1)}) & Cov(X_{1(3)}, X_{1(2)}) & Var(X_{1(3)}) & \cdots & Cov(X_{1(3)}, X_{1(k)}) \\ \vdots & \vdots & \vdots & \ddots & \vdots \\ Cov(X_{1(k)}, X_{1(1)}) & Cov(X_{1(k)}, X_{1(2)}) & Cov(X_{1(k)}, X_{1(3)}) & \cdots & Var(X_{1(k)}) \end{bmatrix}.$$

Central Limit Theorems for Dependent Processes

CLT under weak dependence: A useful generalization of a sequence of independent, identically distributed random variables is a mixing random process in discrete time; "mixing" means, roughly, that random variables temporally far apart from one another are nearly independent. Several kinds of mixing are used in ergodic theory and probability theory.

A simplified formulation of the central limit theorem under strong mixing is:

Theorem. Suppose that $X_1, X_2, \ldots$ is stationary and α-mixing with $\alpha_n = O(n^{-5})$ and that $E(X_n) = 0$ and $E(X_n^{12}) < \infty$. Denote $S_n = X_1 + \ldots + X_n$, then the limit

$$\sigma^2 = \lim_n \frac{E(S_n^2)}{n}$$

exists, and if $\sigma \neq 0$ then $S_n / (\sigma\sqrt{n})$ converges in distribution to $N(0, 1)$.

In fact,

$$\sigma^2 = E(X_1^2) + 2\sum_{k=1}^{\infty} E(X_1 X_{1+k}),$$

where the series converges absolutely. The assumption $\sigma \neq 0$ cannot be omitted, since the asymptotic normality fails for $X_n = Y_n - Y_{n-1}$ where Y_n are another stationary sequence.

There is a stronger version of the theorem: the assumption $E(X_n^{12}) < \infty$ is replaced with $E(|X_n|^{2+\delta}) < \infty$, and the assumption $\alpha_n = O(n^{-5})$ is replaced with $\sum_n \alpha_n^{\frac{\delta}{2(2+\delta)}} < \infty$. Existence of such $\delta > 0$ ensures the conclusion.

Martingale Difference CLT

Theorem. Let a martingale M_n satisfy

- $\frac{1}{n}\sum_{k=1}^{n} E((M_k - M_{k-1})^2 \mid M_1, \ldots, M_{k-1}) \to 1$ in probability as n tends to infinity,
- for every $\varepsilon > 0$, $\frac{1}{n}\sum_{k=1}^{n} E\left((M_k - M_{k-1})^2; |M_k - M_{k-1}| > \varepsilon\sqrt{n}\right) \to 0$ as n tends to infinity, then $M_n / \sqrt{n}$ converges in distribution to $N(0,1)$ as $n \to \infty$.

Caution: The restricted expectation $E(X; A)$ should not be confused with the conditional expectation $E(X|A) = E(X; A)/P(A)$.

Remarks

Proof of classical CLT: For a theorem of such fundamental importance to statistics and applied probability, the central limit theorem has a remarkably simple proof using characteristic functions. It is similar to the proof of a (weak) law of large numbers. For any random variable, Y, with zero mean and a unit variance ($\text{var}(Y) = 1$), the characteristic function of Y is, by Taylor's theorem,

$$\varphi_Y(t) = 1 - \frac{t^2}{2} + o(t^2), \quad t \to 0$$

where $o(t^2)$ is "little o notation" for some function of t that goes to zero more rapidly than t^2. Letting Y_i be $(X_i - \mu)/\sigma$, the standardized value of X_i, it is easy to see that the standardized mean of the observations $X_1, X_2, \ldots, X_n$ is

$$Z_n = \frac{n\bar{X}_n - n\mu}{\sigma\sqrt{n}} = \sum_{i=1}^{n} \frac{Y_i}{\sqrt{n}}.$$

By simple properties of characteristic functions, the characteristic function of Z_n is

$$\left[\varphi_Y\left(\frac{t}{\sqrt{n}}\right)\right]^n = \left[1 - \frac{t^2}{2n} + o\left(\frac{t^2}{n}\right)\right]^n \to e^{-t^2/2}, \quad n \to \infty.$$

But this limit is just the characteristic function of a standard normal distribution N(0, 1), and the central limit theorem follows from the Lévy continuity theorem, which confirms that the convergence of characteristic functions implies convergence in distribution.

Convergence to the Limit

The central limit theorem gives only an asymptotic distribution.

As an approximation for a finite number of observations, it provides a reasonable approximation only when close to the peak of the normal distribution; it requires a very large number of observations to stretch into the tails.

If the third central moment $E((X_1 - \mu)^3)$ exists and is finite, then the above convergence is uniform and the speed of convergence is at least on the order of $1/n^{1/2}$.

Stein's method can be used not only to prove the central limit theorem, but also to provide bounds on the rates of convergence for selected metrics.

The convergence to the normal distribution is monotonic, in the sense that the entropy of Z_n increases monotonically to that of the normal distribution.

The central limit theorem applies in particular to sums of independent and identically distributed discrete random variables. A sum of discrete random variables is still a discrete random variable, so that we are confronted with a sequence of discrete random variables whose cumulative probability distribution function converges towards a cumulative probability distribution function corresponding to a continuous variable (namely that of the normal distribution).

This means that if we build a histogram of the realisations of the sum of n independent identical discrete variables, the curve that joins the centres of the upper faces of the rectangles forming the histogram converges towards a Gaussian curve as n approaches infinity, this relation is known as de Moivre–Laplace theorem. The binomial distribution article details such an application of the central limit theorem in the simple case of a discrete variable taking only two possible values.

Relation to the Law of Large Numbers

The law of large numbers as well as the central limit theorem are partial solutions to a general problem: "What is the limiting behaviour of S_n as n approaches infinity?" In mathematical analysis, asymptotic series are one of the most popular tools employed to approach such questions.

Suppose we have an asymptotic expansion of $f(n)$:

$$f(n) = a_1\varphi_1(n) + a_2\varphi_2(n) + O(\varphi_3(n)) \qquad (n \to \infty).$$

Dividing both parts by $\varphi_1(n)$ and taking the limit will produce a_1, the coefficient of the highest-order term in the expansion, which represents the rate at which $f(n)$ changes in its leading term.

$$\lim_{n\to\infty} \frac{f(n)}{\varphi_1(n)} = a_1.$$

Informally, one can say: "$f(n)$ grows approximately as a_1 $\varphi_1(n)$". Taking the difference between $f(n)$ and its approximation and then dividing by the next term in the expansion, we arrive at a more refined statement about $f(n)$:

$$\lim_{n\to\infty}\frac{f(n)-a_1\varphi_1(n)}{\varphi_2(n)}=a_2.$$

Here one can say that the difference between the function and its approximation grows approximately as $a_2\ \varphi_2(n)$. The idea is that dividing the function by appropriate normalizing functions, and looking at the limiting behaviour of the result, can tell us much about the limiting behaviour of the original function itself.

Informally, something along these lines is happening when the sum, S_n, of independent identically distributed random variables, X_1, ..., X_n, is studied in classical probability theory. If each X_i has finite mean μ, then by the law of large numbers, $S_n/n \to \mu$. If in addition each X_i has finite variance σ^2, then by the central limit theorem,

$$\frac{S_n - n\mu}{\sqrt{n}} \to \xi,$$

where ∂ is distributed as $N(0, \sigma^2)$. This provides values of the first two constants in the informal expansion

$$S_n \approx \mu n + \xi\sqrt{n}.$$

In the case where the X_i's do not have finite mean or variance, convergence of the shifted and rescaled sum can also occur with different centering and scaling factors:

$$\frac{S_n - a_n}{b_n} \to \Xi,$$

or informally

$$S_n \approx a_n + \Xi b_n.$$

Distributions ∂ which can arise in this way are called *stable*. Clearly, the normal distribution is stable, but there are also other stable distributions, such as the Cauchy distribution, for which the mean or variance are not defined. The scaling factor b_n may be proportional to n^c, for any $c \geq 1/2$; it may also be multiplied by a slowly varying function of n.

The law of the iterated logarithm specifies what is happening "in between" the law of large numbers and the central limit theorem. Specifically it says that the normalizing function $\sqrt{n \log \log n}$ intermediate in size between n of the law of large numbers and $\sqrt{n}$ of the central limit theorem provides a non-trivial limiting behaviour.

Alternative Statements of the Theorem

Density functions: The density of the sum of two or more independent variables is the convolution of their densities (if these densities exist). Thus the central limit theorem can be interpreted as a statement about the properties of density functions under convolution: the convolution of a number of density functions tends to the normal density as the number of density functions increases without bound. These theorems require stronger hypotheses than the forms of the central limit theorem given above. Theorems of this type are often called local limit theorems.

Characteristic Functions

Since the characteristic function of a convolution is the product of the characteristic functions of the densities involved, the central limit theorem has yet another restatement: the product of the characteristic functions of a number of density functions becomes close to the characteristic function of the normal density as the number of density functions increases without bound, under the conditions stated above. However, to state this more precisely, an appropriate scaling factor needs to be applied to the argument of the characteristic function.

An equivalent statement can be made about Fourier transforms, since the characteristic function is essentially a Fourier transform.

Extensions to the Theorem

Products of positive random variables:

The logarithm of a product is simply the sum of the logarithms of the factors. Therefore when the logarithm of a product of random variables that take only positive values approaches a normal distribution, the product itself approaches a log-normal distribution. Many physical quantities (especially mass or length, which are a matter of scale and cannot be negative) are the products of different random factors, so they follow a log-normal distribution.

Whereas the central limit theorem for sums of random variables requires the condition of finite variance, the corresponding theorem for products requires the corresponding condition that the density function be square-integrable.

Beyond the Classical Framework

Asymptotic normality, that is, convergence to the normal distribution after appropriate shift and rescaling, is a phenomenon much more general than the classical framework treated above, namely, sums of independent random variables (or vectors). New frameworks are revealed from time to time; no single unifying framework is available for now.

Convex Body

Theorem. There exists a sequence $\varepsilon_n \downarrow 0$ for which the following holds. Let $n \geq 1$, and let random variables $X_1, \ldots, X_n$ have a log-concave joint density f such that $f(x_1, \ldots, x_n) = f(|x_1|, \ldots, |x_n|)$ for all $x_1, \ldots, x_n$, and $E(X_k^2) = 1$ for all $k = 1, \ldots, n$. Then the distribution of

$$\frac{X_1 + \ldots + X_n}{\sqrt{n}}$$

is ε_n-close to N(0, 1) in the total variation distance.

These two ε_n-close distributions have densities (in fact, log-concave densities), thus, the total variance distance between them is the integral of the absolute value of the difference between the densities. Convergence in total variation is stronger than weak convergence.

An important example of a log-concave density is a function constant inside a given convex body and vanishing outside; it corresponds to the uniform distribution on the convex body, which explains the term "central limit theorem for convex bodies".

Another example: $f(x_1, \ldots, x_n) = \text{const} \cdot \exp(-(|x_1|^\alpha + \ldots + |x_n|^\alpha)^\beta)$ where $\alpha > 1$ and $\alpha\beta > 1$. If $\beta = 1$ then $f(x_1, \ldots, x_n)$ factorizes into const $\cdot \exp(-|x_1|^\alpha) \ldots \exp(-|x_n|^\alpha)$, which means independence of $X_1, \ldots, X_n$. In general, however, they are dependent.

The condition $f(x_1, \ldots, x_n) = f(|x_1|, \ldots, |x_n|)$ ensures that $X_1, \ldots, X_n$ are of zero mean and uncorrelated; still, they need not be independent, nor even pairwise independent. By the way, pairwise independence cannot replace independence in the classical central limit theorem.

Here is a Berry-Esseen Type Result.

Theorem. Let $X_1, \ldots, X_n$ satisfy the assumptions of the previous theorem, then

$$\left| \mathbb{P}\left(a \le \frac{X_1 + \ldots + X_n}{\sqrt{n}} \le b\right) - \frac{1}{\sqrt{2\pi}} \int_a^b e^{-t^2/2} \, dt \right| \le \frac{C}{n}$$

for all $a < b$; here C is a universal (absolute) constant. Moreover, for every $c_1, \ldots, c_n \in \mathrm{R}$ such that $c_1^2 + \ldots + c_n^2 = 1$,

$$\left| \mathbb{P}(a \le c_1 X_1 + \ldots + c_n X_n \le b) - \frac{1}{\sqrt{2\pi}} \int_a^b e^{-t^2/2} \, dt \right| \le C(c_1^4 + \ldots + c_n^4).$$

The distribution of $(X_1 + \ldots + X_n)/\sqrt{n}$ need not be approximately normal (in fact, it can be uniform). However, the distribution of $c_1X_1 + \ldots + c_nX_n$ is close to $N(0, 1)$ (in the total variation distance) for most of vectors $(c_1, \ldots, c_n)$ according to the uniform distribution on the sphere $c_1^2 + \ldots + c_n^2 = 1$.

Lacunary Trigonometric Series

Theorem (Salem - Zygmund). Let U be a random variable distributed uniformly on $(0, 2\pi)$, and $X_k = r_k \cos(n_k U + a_k)$, where

- n_k satisfy the lacunarity condition: there exists $q > 1$ such that $n_{k+1} \ge q n_k$ for all k,
- r_k are such that

$$r_1^2 + r_2^2 + \cdots = \infty \text{ and } \frac{r_k^2}{r_1^2 + \cdots + r_k^2} \to 0,$$

- $0 \le a_k < 2\pi$. Then

$$\frac{X_1 + \cdots + X_k}{\sqrt{r_1^2 + \cdots + r_k^2}}$$

converges in distribution to $N(0, 1/2)$.

Gaussian Polytopes

Theorem Let $A_1, \ldots, A_n$ be independent random points on the plane R^2 each having the two-dimensional standard normal distribution. Let K_n be the convex hull of these points, and X_n the area of K_n Then

$$\frac{X_n - EX_n}{\sqrt{\operatorname{Var} X_n}}$$

converges in distribution to $N(0, 1)$ as n tends to infinity.

The same holds in all dimensions (2, 3, ...).

The polytope K_n is called Gaussian random polytope.

A similar result holds for the number of vertices (of the Gaussian polytope), the number of edges, and in fact, faces of all dimensions.

Linear Functions of Orthogonal Matrices

A linear function of a matrix M is a linear combination of its elements (with given coefficients), $M \mapsto \operatorname{tr}(AM)$ where A is the matrix of the coefficients.

A random orthogonal matrix is said to be distributed uniformly, if its distribution is the normalized Haar measure on the orthogonal group $O(n, R)$.

Theorem. Let M be a random orthogonal $n \times n$ matrix distributed uniformly, and A a fixed $n \times n$ matrix such that $\operatorname{tr}(AA^*) = n$, and let $X = \operatorname{tr}(AM)$. Then the distribution of X is close to $N(0, 1)$ in the total variation metric up to $2\sqrt{3}/(n-1)$.

Subsequences

Theorem. Let random variables $X_1, X_2, \ldots \in L_2(\Omega)$ be such that $X_n \to 0$ weakly in $L_2(\Omega)$ and $X_n^2 \to 1$ weakly in $L_1(\Omega)$. Then there exist integers $n_1 < n_2 < \ldots$ such that $(X_{n_1} + \cdots + X_{n_k}) / \sqrt{k}$ converges in distribution to $N(0, 1)$ as k tends to infinity.

Q-analogues

A generalized q-analog of the classical central limit theorem has been described by Umarov, Tsallis and Steinberg in which the independence constraint for the i.i.d. variables is relaxed to an extent defined by the q parameter, with independence being recovered as q->1. In analogy to the classical central limit theorem, such random variables with fixed mean and variance tend towards the q-Gaussian distribution, which maximizes the Tsallis entropy under these constraints. Umarov, Tsallis, Gell-Mann and Steinberg have defined q-analogs of all symmetric alpha-stable distributions, and have formulated a number of conjectures regarding their relevance to an even more general Central limit theorem.

Applications and Examples

Simple Example

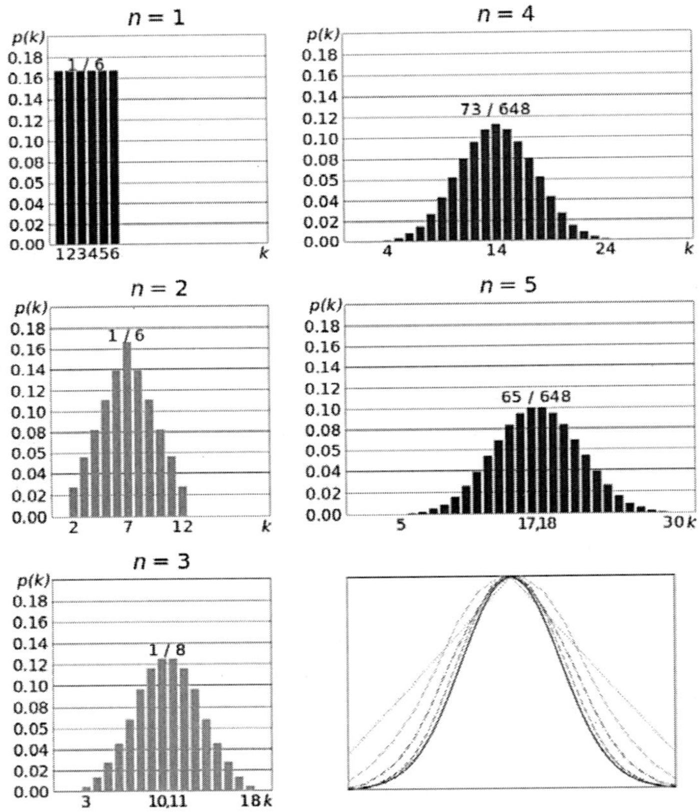

Figure: *Comparison of probability density functions,* p(k) *for the sum of* n *fair 6-sided dice to show their convergence to a normal distribution with increasing* n, *in accordance to the central limit theorem. In the bottom-right graph, smoothed profiles of the previous graphs are rescaled, superimposed and compared with a normal distribution (black curve).*

A simple example of the central limit theorem is rolling a large number of identical, unbiased dice. The distribution of the sum (or average) of the rolled numbers will be well approximated by a normal distribution. Since real-world quantities are often the balanced sum of many unobserved random events, the central limit theorem also provides a partial explanation for the prevalence of the normal probability distribution. It also justifies the approximation of large-sample statistics to the normal distribution in controlled experiments.

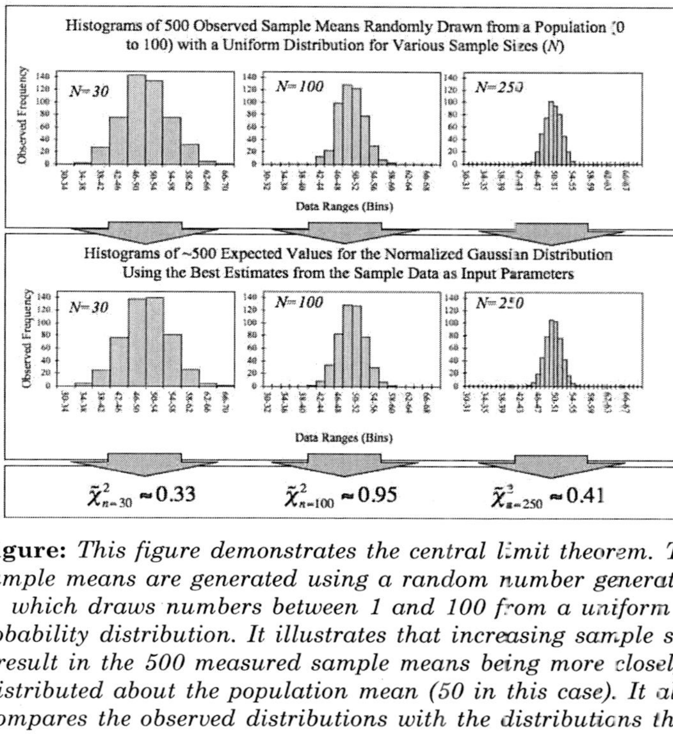

Figure: *This figure demonstrates the central limit theorem. The sample means are generated using a random number generator, which draws numbers between 1 and 100 from a uniform probability distribution. It illustrates that increasing sample sizes result in the 500 measured sample means being more closely distributed about the population mean (50 in this case). It also compares the observed distributions with the distributions that would be expected for a normalized Gaussian distribution, and shows the chi-squared values that quantify the goodness of the fit (the fit is good if the reduced chi-squared value is less than or approximately equal to one). The input into the normalized Gaussian function is the mean of sample means (~50) and the mean sample standard deviation divided by the square root of the sample size (~28.87/ $\sqrt{n}$), which is called the standard deviation of the mean (since it refers to the spread of sample means).*

Real Applications

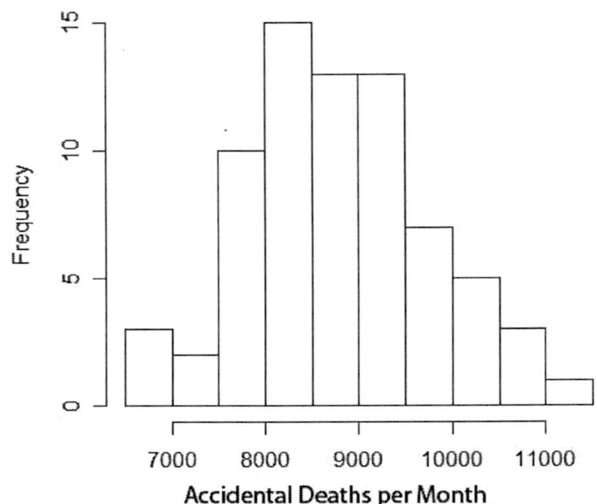

Figure: *A histogram plot of monthly accidental deaths in the US, between 1973 and 1978 exhibits normality, due to the central limit theorem*

Published literature contains a number of useful and interesting examples and applications relating to the central limit theorem. One source states the following examples:

- The probability distribution for total distance covered in a random walk (biased or unbiased) will tend towards a normal distribution.
- Flipping a large number of coins will result in a normal distribution for the total number of heads (or equivalently total number of tails).

From another viewpoint, the central limit theorem explains the common appearance of the "Bell Curve" in density estimates applied to real world data. In cases like electronic noise, examination grades, and so on, we can often regard a single measured value as the weighted average of a large number of small effects. Using generalisations of the central limit theorem, we can then see that this would often (though not always) produce a final distribution that is approximately normal.

In general, the more a measurement is like the sum of independent variables with equal influence on the result, the more normality it exhibits. This justifies the common use of this distribution to stand in for the effects of unobserved variables in models like the linear model.

Other Illustrations

Given its importance to statistics, a number of papers and computer packages are available that demonstrate the convergence involved in the central limit theorem.

Applications

Probability theory is applied in everyday life in risk assessment and in trade on financial markets. Governments apply probabilistic methods in environmental regulation, where it is called pathway analysis. A good example is the effect of the perceived probability of any widespread Middle East conflict on oil prices—which have ripple effects in the economy as a whole. An assessment by a commodity trader that a war is more likely vs. less likely sends prices up or down, and signals other traders of that opinion. Accordingly, the probabilities are neither assessed independently nor necessarily very rationally. The theory of behavioural finance emerged to describe the effect of such groupthink on pricing, on policy, and on peace and conflict.

The discovery of rigorous methods to assess and combine probability assessments has changed society. It is important for most citizens to understand how probability assessments are made, and how they contribute to decisions.

Another significant application of probability theory in everyday life is reliability. Many consumer products, such as automobiles and consumer electronics, use reliability theory in product design to reduce the probability of failure. Failure probability may influence a manufacture's decisions on a product's warranty.

The cache language model and other statistical language models that are used in natural language processing are also examples of applications of probability theory.

Mathematical Treatment

Consider an experiment that can produce a number of results. The collection of all results is called the sample space of the experiment. The power set of the sample space is formed by considering all different collections of possible results. For example, rolling a die can produce six possible results. One collection of possible results gives an odd number on the die. Thus, the subset {1,3,5} is an element of the power set of the sample space of die rolls.

These collections are called "events." In this case, {1,3,5} is the event that the die falls on some odd number. If the results that actually occur fall in a given event, the event is said to have occurred.

A probability is a way of assigning every event a value between zero and one, with the requirement that the event made up of all possible results (in our example, the event {1,2,3,4,5,6}) is assigned a value of one. To qualify as a probability, the assignment of values must satisfy the requirement that if you look at a collection of mutually exclusive events (events with no common results, e.g., the events {1,6}, {3}, and {2,4} are all mutually exclusive), the probability that at least one of the events will occur is given by the sum of the probabilities of all the individual events.

The probability of an event A is written as P(A), p(A) or Pr(A). This mathematical definition of probability can extend to infinite sample spaces, and even uncountable sample spaces, using the concept of a measure.

The *opposite* or *complement* of an event A is the event [not A] (that is, the event of A not occurring); its probability is given by P(not A) = 1 - P(A). As an example, the chance of not rolling a six on a six-sided die is $1 -$ (chance of rolling a six) $= 1 - \frac{1}{6} = \frac{5}{6}$.

If both events A and B occur on a single performance of an experiment, this is called the intersection or joint probability of A and B, denoted as $P(A \cap B)$.

Independent Probability

If two events, A and B are independent then the joint probability is

$$P(A \text{ and } B) = P(A \cap B) = P(A)P(B),$$

for example, if two coins are flipped the chance of both being heads is $\frac{1}{2} \times \frac{1}{2} = \frac{1}{4}$.

Mutually Exclusive

If either event A or event B or both events occur on a single performance of an experiment this is called the union of the events A and B denoted as $P(A \cup B)$. If two events are mutually exclusive then the probability of either occurring is

$$P(A \text{ or } B) = P(A \cup B) = P(A) + P(B).$$

For example, the chance of rolling a 1 or 2 on a six-sided die is $P(1 \text{ or } 2) = P(1) + P(2) = \frac{1}{6} + \frac{1}{6} = \frac{1}{3}$.

Not Mutually Exclusive

If the events are not mutually exclusive then

$$\mathrm{P}(A \text{ or } B) = \mathrm{P}(A) + \mathrm{P}(B) - \mathrm{P}(A \text{ and } B).$$

For example, when drawing a single card at random from a regular deck of cards, the chance of getting a heart or a face card (J,Q,K) (or one that is both) is $\frac{13}{52} + \frac{12}{52} - \frac{3}{52} = \frac{11}{26}$, because of the 52 cards of a deck 13 are hearts, 12 are face cards, and 3 are both: here the possibilities included in the "3 that are both" are included in each of the "13 hearts" and the "12 face cards" but should only be counted once.

Conditional Probability

Conditional probability is the probability of some event A, given the occurrence of some other event B. Conditional probability is written $\mathrm{P}(A \mid B)$, and is read "the probability of A, given B". It is defined by

$$\mathrm{P}(A \mid B) = \frac{\mathrm{P}(A \cap B)}{\mathrm{P}(B)}.$$

If $\mathrm{P}(B) = 0$ then $\mathrm{P}(A \mid B)$ is formally undefined by this expression. However, it is possible to define a conditional probability for some zero-probability events using a σ-algebra of such events (such as those arising from a continuous random variable).

For example, in a bag of 2 red balls and 2 blue balls (4 balls in total), the probability of taking a red ball is $1/2$; however, when taking a second ball, the probability of it being either a red ball or a blue ball depends on the ball previously taken, such as, if a red ball was taken, the probability of picking a red ball again would be $1/3$ since only 1 red and 2 blue balls would have been remaining.

Summary of Probabilities

Summary of probabilities

Event Probability

A $P(A) \in [0,1]$

not A $P(A^c) = 1 - P(A)$

A or B $P(A \cup B) = P(A) + P(B) - P(A \cap B)$
$P(A \cup B) = P(A) + P(B)$ if A and B are mutually exclusive

A and B $P(A \cap B) = P(A \mid B)P(B) = P(B \mid A)P(A)$
$P(A \cap B) = P(A)P(B)$ if A and B are independent

A given B $P(A \mid B) = \dfrac{P(A \cap B)}{P(B)} = \dfrac{P(B \mid A)P(A)}{P(B)}$

Graph Theory

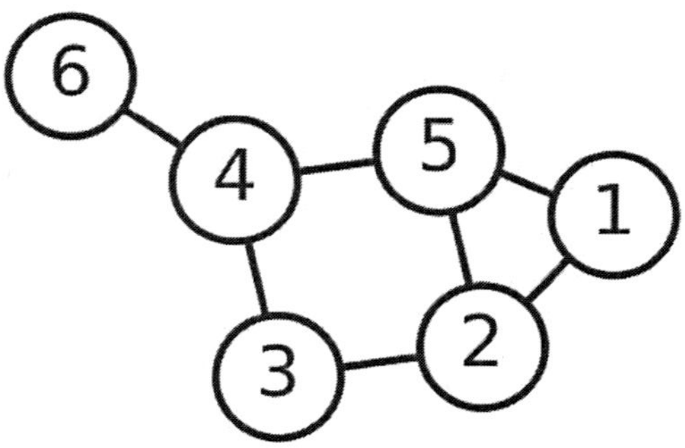

Figure: *A drawing of a graph*

In mathematics and computer science, graph theory is the study of *graphs*, which are mathematical structures used to model pairwise relations between objects. A *graph* in this context is made up of *vertices* or *nodes* and lines called *edges* that connect them. A graph may be *undirected*, meaning that there is no distinction between the two vertices associated with each edge, or its edges may be *directed* from one vertex to another.

The graphs studied in graph theory should not be confused with the graphs of functions or other kinds of graphs.

Refer to the glossary of graph theory for basic definitions in graph theory.

Applications

Graphs can be used to model many types of relations and process dynamics in physical, biological, social and information systems. Many practical problems can be represented by graphs.

In computer science, graphs are used to represent networks of communication, data organization, computational devices, the flow of computation, etc. One practical example: The link structure of a website could be represented by a directed graph. The vertices are the web pages available at the website and a directed edge from page A to page B exists if and only if A contains a link to B.

A similar approach can be taken to problems in travel, biology, computer chip design, and many other fields. The development of algorithms to handle graphs is therefore of major interest in computer science. There, the transformation of graphs is often formalized and represented by graph rewrite systems. They are either directly used or properties of the rewrite systems (e.g. confluence) are studied. Complementary to graph transformation systems focussing on rule-based in-memory manipulation of graphs are graph databases geared towards transaction-safe, persistent storing and querying of graph-structured data.

Graph-theoretic methods, in various forms, have proven particularly useful in linguistics, since natural language often lends itself well to discrete structure. Traditionally, syntax and compositional semantics follow tree-based structures, whose expressive power lies in the Principle of Compositionality, modelled in a hierarchical graph.

More contemporary approaches such as Head-driven phrase structure grammar (HPSG) model syntactic constructions via the unification of typed feature structures, which are directed acyclic graphs. Within lexical semantics, especially as applied to computers, modelling word meaning is easier when a given word is understood in terms of related words; semantic networks are therefore important in computational linguistics. Still other methods in phonology (e.g. Optimality Theory, which uses lattice graphs) and morphology (e.g. finite-state morphology, using finite-state transducers) are common in the analysis of language as a graph. Indeed, the usefulness of this area of mathematics to linguistics has borne organizations such as TextGraphs, as well as various 'Net' projects, such as WordNet, VerbNet, and others.

Graph theory is also used to study molecules in chemistry and physics. In condensed matter physics, the three dimensional structure of complicated simulated atomic structures can be studied quantitatively by gathering statistics on graph-theoretic properties related to the topology of the atoms. For example, Franzblau's shortest-path (SP) rings. In chemistry a graph makes a natural model for a molecule, where vertices represent atoms and edges bonds. This approach is especially used in computer processing of molecular structures, ranging from chemical editors to database searching. In statistical physics, graphs can represent local connections between interacting parts of a system, as well as the dynamics of a physical process on such systems.

Graph theory is also widely used in sociology as a way, for example, to measure actors' prestige or to explore diffusion mechanisms, notably through the use of social network analysis software. Under the umbrella of Social Network graphs there are many different types of graphs: Starting with the Acquaintanceship and Friendship Graphs, these graphs are useful for representing whether n people know each other. next there is the influence graph. This graph is used to model whether certain people can influence the behaviour of others. Finally there's a collaboration graph which models whether two people work together in a particular way. The measure of an actors' prestige mentioned above is an example of this, other popular examples include the Erdõs number and six degrees of separation

Likewise, graph theory is useful in biology and conservation efforts where a vertex can represent regions where certain species exist (or habitats) and the edges represent migration paths, or movement between the regions. This information is important when looking at breeding patterns or tracking the spread of disease, parasites or how changes to the movement can affect other species.

In mathematics, graphs are useful in geometry and certain parts of topology, e.g. Knot Theory. Algebraic graph theory has close links with group theory.

A graph structure can be extended by assigning a weight to each edge of the graph. Graphs with weights, or weighted graphs, are used to represent structures in which pairwise connections have some numerical values. For example if a graph represents a road network, the weights could represent the length of each road.

A digraph with weighted edges in the context of graph theory is called a network. Network analysis have many practical applications, for example, to model and analyze traffic networks. Applications of network analysis split broadly into three categories:

1. First, analysis to determine structural properties of a network, such as the distribution of vertex degrees and the diameter of the graph. A vast number of graph measures exist, and the production of useful ones for various domains remains an active area of research.
2. Second, analysis to find a measurable quantity within the network, for example, for a transportation network, the level of vehicular flow within any portion of it.
3. Third, analysis of dynamical properties of networks.

History

The paper written by Leonhard Euler on the *Seven Bridges of Königsberg* and published in 1736 is regarded as the first paper in the history of graph theory. This paper, as well as the one written by Vandermonde on the *knight problem,* carried on with the *analysis situs* initiated by Leibniz. Euler's formula relating the number of edges, vertices, and faces of a convex polyhedron was studied and generalized by Cauchy and L'Huillier, and is at the origin of topology.

More than one century after Euler's paper on the bridges of Königsberg and while Listing introduced topology, Cayley was led by the study of particular analytical forms arising from differential calculus to study a particular class of graphs, the *trees*. This study had many implications in theoretical chemistry. The involved techniques mainly concerned the enumeration of graphs having particular properties.

Enumerative graph theory then rose from the results of Cayley and the fundamental results published by Pólya between 1935 and 1937 and the generalization of these by De Bruijn in 1959. Cayley linked his results on trees with the contemporary studies of chemical composition. The fusion of the ideas coming from mathematics with those coming from chemistry is at the origin of a part of the standard terminology of graph theory.

In particular, the term "graph" was introduced by Sylvester in a paper published in 1878 in *Nature*, where he draws an analogy between "quantic invariants" and "co-variants" of algebra and molecular diagrams:

"[...] Every invariant and co-variant thus becomes expressible by a *graph* precisely identical with a Kekuléan diagram or chemicograph. [...] I give a rule for the geometrical multiplication of graphs, *i.e.* for constructing a *graph* to the product of in- or co-variants whose separate graphs are given. [...]" (italics as in the original).

The first textbook on graph theory was written by Dénes Kõnig, and published in 1936. A later textbook by Frank Harary, published in 1969, was enormously popular, and enabled mathematicians, chemists, electrical engineers and social scientists to talk to each other. Harary donated all of the royalties to fund the Pólya Prize.

One of the most famous and productive problems of graph theory is the four colour problem: "Is it true that any map drawn in the plane may have its regions coloured with four colours, in such a way that any two regions having a common border have different colours?"

This problem was first posed by Francis Guthrie in 1852 and its first written record is in a letter of De Morgan addressed to Hamilton the same year. Many incorrect proofs have been proposed, including those by Cayley, Kempe, and others. The study and the generalization of this problem by Tait, Heawood, Ramsey and Hadwiger led to the study of the colourings of the graphs embedded on surfaces with arbitrary genus. Tait's reformulation generated a new class of problems, the *factorization problems*, particularly studied by Petersen and Kõnig. The works of Ramsey on colourations and more specially the results obtained by Turán in 1941 was at the origin of another branch of graph theory, *extremal graph theory*.

The four colour problem remained unsolved for more than a century. In 1969 Heinrich Heesch published a method for solving the problem using computers. A computer-aided proof produced in 1976 by Kenneth Appel and Wolfgang Haken makes fundamental use of the notion of "discharging" developed by Heesch. The proof involved checking the properties of 1,936 configurations by computer, and was not fully accepted at the time due to its complexity. A simpler proof considering only 633 configurations was given twenty years later by Robertson, Seymour, Sanders and Thomas.

The autonomous development of topology from 1860 and 1930 fertilized graph theory back through the works of Jordan, Kuratowski and Whitney. Another important factor of common development of graph theory and topology came from the use of the techniques of modern algebra. The first example of such a use comes from the work of the physicist Gustav Kirchhoff, who published in 1845 his Kirchhoff's circuit laws for calculating the voltage and current in electric circuits.

The introduction of probabilistic methods in graph theory, especially in the study of Erdõs and Rényi of the asymptotic probability of graph connectivity, gave rise to yet another branch, known as *random graph theory*, which has been a fruitful source of graph-theoretic results.

Graph Drawing

Graph drawing is an area of mathematics and computer science combining methods from geometric graph theory and information visualization to derive two-dimensional depictions of graphs arising from applications such as social network analysis, cartography, and bioinformatics.

A drawing of a graph or network diagram is a pictorial representation of the vertices and edges of a graph. This drawing should not be confused with the graph itself: very different layouts can correspond to the same graph. In the abstract, all that matters is which pairs vertices are connected by edges. In the concrete, however, the arrangement of these vertices and edges within a drawing affects its understandability, usability, fabrication cost, and aesthetics. The problem gets worse, if the graph changes over time by adding and deleting edges (dynamic graph drawing) and the goal is to preserve the user's mental map.

Graphical Conventions

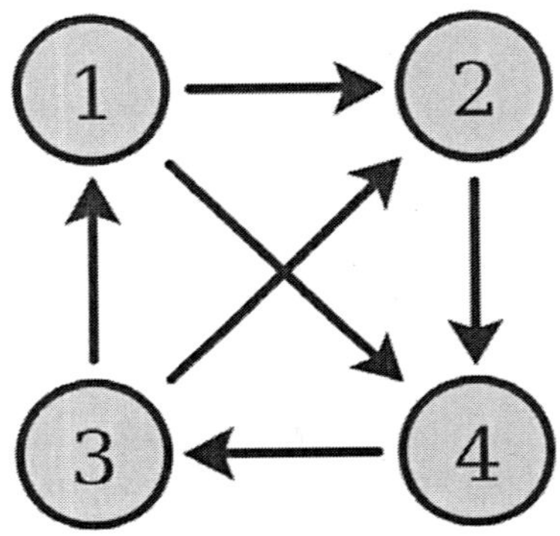

***Figure:** A directed graph with arrowheads showing the edge directions*

Graphs are frequently drawn as node-link diagrams in which the vertices are represented as disks or boxes and the edges are represented as line segments, polylines, or curves in the Euclidean plane.

In the case of directed graphs, arrowheads form a commonly used graphical convention to show their orientation; however, user studies have shown that other conventions such as tapering provide this information more effectively.

Alternative conventions to node-link diagrams include adjacency representations such as circle packings, in which vertices are represented by disjoint regions in the plane and edges are represented by adjacencies between regions; intersection representations in which vertices are represented by non-disjoint geometric objects and edges are represented by their intersections; visibility representations in which vertices are represented by regions in the plane and edges are represented by regions that have an unobstructed line of sight to each other; confluent drawings, in which edges are represented as smooth curves within mathematical train tracks; and visualizations of the adjacency matrix of the graph.

Quality Measures

Many different quality measures have been defined for graph drawings, in an attempt to find objective means of evaluating their aesthetics and usability. In addition to guiding the choice between different layout methods for the same graph, some layout methods attempt to directly optimize these measures.

- The crossing number of a drawing is the number of pairs of edges that cross each other. If the graph is planar, then it is often convenient to draw it without any edge intersections; that is, in this case, a graph drawing represents a graph embedding. However, nonplanar graphs frequently arise in applications, so graph drawing algorithms must generally allow for edge crossings.
- The area of a drawing is the size of its smallest bounding box, relative to the closest distance between any two vertices. Drawings with smaller area are generally preferable to those with larger area, because they allow the features of the drawing to be shown at greater size and therefore more legibly. The aspect ratio of the bounding box may also be important.
- Symmetry display is the problem of finding symmetry groups within a given graph, and finding a drawing that displays as much of the symmetry as possible. Some layout methods automatically lead to symmetric drawings; alternatively, some drawing methods start by finding symmetries in the input graph and using them to construct a drawing.
- It is important that edges have shapes that are as simple as possible, to make it easier for the eye to follow them. In polyline drawings, the complexity of an edge may be measured by its number of bends, and many methods aim to provide drawings with few total bends or few bends per edge. Similarly for spline curves the complexity of an edge may be measured by the number of control points on the edge.

- Several commonly used quality measures concern lengths of edges: it is generally desirable to minimize the total length of the edges as well as the maximum length of any edge. Additionally, it may be preferable for the lengths of edges to be uniform rather than highly varied.
- Angular resolution is a measure of the sharpest angles in a graph drawing. If a graph has vertices with high degree then it necessarily will have small angular resolution, but the angular resolution can be bounded below by a function of the degree.
- The slope number of a graph is the minimum number of distinct edge slopes needed in a drawing with straight line segment edges (allowing crossings). Cubic graphs have slope number at most four, but graphs of degree five may have unbounded slope number; it remains open whether the slope number of degree-4 graphs is bounded.

Layout Methods

There are many different graph layout strategies:

- In force-based layout systems, the graph drawing software modifies an initial vertex placement by continuously moving the vertices according to a system of forces based on physical metaphors related to systems of springs or molecular mechanics. Typically, these systems combine attractive forces between adjacent vertices with repulsive forces between all pairs of vertices, in order to seek a layout in which edge lengths are small while vertices are well-separated. These systems may perform gradient descent based minimization of an energy function, or they may translate the forces directly into velocities or accelerations for the moving vertices.
- Spectral layout methods use as coordinates the eigenvectors of a matrix such as the Laplacian derived from the adjacency matrix of the graph.
- Orthogonal layout methods, which allow the edges of the graph to run horizontally or vertically, parallel to the coordinate axes of the layout. These methods were

originally designed for VLSI and PCB layout problems but they have also been adapted for graph drawing. They typically involve a multiphase approach in which an input graph is planarized by replacing crossing points by vertices, a topological embedding of the planarized graph is found, edge orientations are chosen to minimize bends, vertices are placed consistently with these orientations, and finally a layout compaction stage reduces the area of the drawing.

- Tree layout algorithms these show a rooted tree-like formation, suitable for trees. Often, in a technique called "balloon layout", the children of each node in the tree are drawn on a circle surrounding the node, with the radii of these circles diminishing at lower levels in the tree so that these circles do not overlap.
- Layered graph drawing methods (often called Sugiyama-style drawing) are best suited for directed acyclic graphs or graphs that are nearly acyclic, such as the graphs of dependencies between modules or functions in a software system. In these methods, the nodes of the graph are arranged into horizontal layers using methods such as the Coffman–Graham algorithm, in such a way that most edges go downwards from one layer to the next; after this step, the nodes within each layer are arranged in order to minimize crossings.

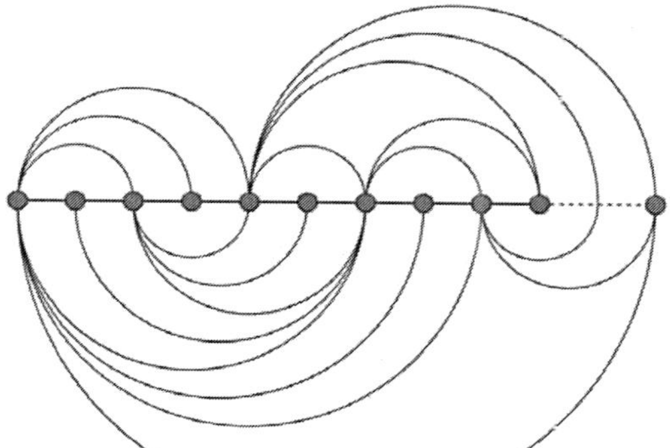

Figure: *Linear embedding*

- Linear embedding places vertices on a line; edges may be drawn as semicircles above or below the line, or as smooth curves linked together from multiple semicircles.
- Circular layout methods place the vertices of the graph on a circle, choosing carefully the ordering of the vertices around the circle to reduce crossings and place adjacent vertices close to each other. Edges may be drawn either as chords of the circle or as arcs inside or outside of the circle. In some cases, multiple circles may be used.

Application-specific Graph Drawings

Graphs and graph drawings arising in other areas of application include

- Sociograms, drawings of a social network, as often offered by social network analysis software
- Hasse diagrams, a type of graph drawing specialized to partial orders
- Dessin d'enfants, a type of graph drawing used in algebraic geometry
- State diagrams, graphical representations of finite state machines
- Computer network diagrams, depictions of the nodes and connections in a computer network.
- Flow charts, drawings in which the nodes represent the steps of an algorithm and the edges represent control flow between steps.
- Data flow diagrams, drawings in which the nodes represent the components of an information system and the edges represent the movement of information from one component to another.

In addition, the placement and routing steps of electronic design automation are similar in many ways to graph drawing, and the graph drawing literature includes several results borrowed from the EDA literature.

However, these problems also differ in several important ways: for instance, in EDA, area minimization and signal length are more important than aesthetics, and the routing problem in EDA may have more than two terminals per net while the analogous problem in graph drawing generally only involves pairs of vertices for each edge.

Graph (Abstract Data Type)

Operations

The basic operations provided by a graph data structure G usually include:

- adjacent(G, x, y): tests whether there is an edge from node x to node y.
- neighbours(G, x): lists all nodes y such that there is an edge from x to y.
- add(G, x, y): adds to G the edge from x to y, if it is not there.
- delete(G, x, y): removes the edge from x to y, if it is there.
- get_node_value(G, x): returns the value associated with the node x.
- set_node_value(G, x, a): sets the value associated with the node x to a.

Structures that associate values to the edges usually also provide:

- get_edge_value(G, x, y): returns the value associated to the edge (x,y).
- set_edge_value(G, x, y, v): sets the value associated to the edge (x,y) to v.

Representations

Different data structures for the representation of graphs are used in practice:

Adjacency List

In graph theory and computer science, an adjacency list representation of a graph is a collection of unordered lists, one for each vertex in the graph. Each list describes the set of neighbours of its vertex.

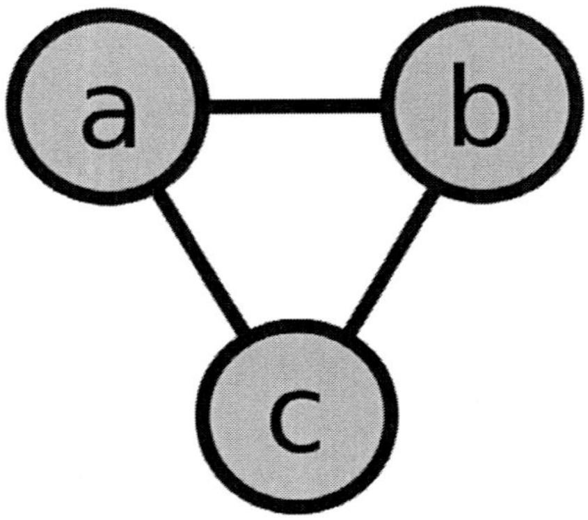

Figure: *This undirected cyclic graph can be described by the list {a,b}, {a,c}, {b,c}.*

Implementation Details

***The graph pictured above has this adjacency list representation*:**		
a	adjacent to	b,c
b	adjacent to	a,c
c	adjacent to	a,b

An adjacency list representation for a graph associates each vertex in the graph with the collection of its neighbouring vertices or edges. There are many variations of this basic idea, differing in the details of how they implement the association between vertices and collections, in how they implement the collections, in whether they include both vertices and edges or only vertices as first class objects, and in what kinds of objects are used to represent the vertices and edges.

- An implementation suggested by Guido van Rossum uses a hash table to associate each vertex in a graph with an array of adjacent vertices. In this representation, a vertex may be represented by any hashable object. There is no explicit representation of edges as objects.
- Cormen et al. suggest an implementation in which the vertices are represented by index numbers. Their

representation uses an array indexed by vertex number, in which the array cell for each vertex points to a singly linked list of the neighbouring vertices of that vertex. In this representation, the nodes of the singly linked list may be interpreted as edge objects; however, they do not store the full information about each edge (they only store one of the two endpoints of the edge) and in undirected graphs there will be two different linked list nodes for each edge (one within the lists for each of the two endpoints of the edge).

- The object oriented incidence list structure suggested by Goodrich and Tamassia has special classes of vertex objects and edge objects. Each vertex object has an instance variable pointing to a collection object that lists the neighbouring edge object. In turn, each edge object points to the two vertex objects at its endpoints. This version of the adjacency list uses more memory than the version in which adjacent vertices are listed directly, but the existence of explicit edge objects allows it extra flexibility in storing additional information about edges.

Operations

The main operation performed by the adjacency list data structure is to report a list of the neighbours of a given vertex. Using any of the implementations detailed above, this can be performed in constant time per neighbour. In other words, the total time to report all of the neighbours of a vertex v is proportional to the degree of v.

It is also possible, but not as efficient, to use adjacency lists to test whether an edge exists or does not exist between two specified vertices.

In an adjacency list in which the neighbours of each vertex are unsorted, testing for the existence of an edge may be performed in time proportional to the degree of one of the two given vertices, by using a sequential search through the neighbours of this vertex. If the neighbours are represented as a sorted array, binary search may be used instead, taking time proportional to the logarithm of the degree.

Trade-offs

The main alternative to the adjacency list is the adjacency matrix, a matrix whose rows and columns are indexed by vertices and whose cells contain a Boolean value that indicates whether an edge is present between the vertices corresponding to the row and column of the cell. For a sparse graph (one in which most pairs of vertices are not connected by edges) an adjacency list is significantly more space-efficient than an adjacency matrix (stored as an array): the space usage of the adjacency list is proportional to the number of edges and vertices in the graph, while for an adjacency matrix stored in this way the space is proportional to the square of the number of vertices. However, it is possible to store adjacency matrices more space-efficiently, matching the linear space usage of an adjacency list, by using a hash table indexed by pairs of vertices rather than an array.

The other significant difference between adjacency lists and adjacency matrices is in the efficiency of the operations they perform. In an adjacency list, the neighbours of each vertex may be listed efficiently, in time proportional to the degree of the vertex. In an adjacency matrix, this operation takes time proportional to the number of vertices in the graph, which may be significantly higher than the degree. On the other hand, the adjacency matrix allows testing whether two vertices are adjacent to each other, in constant time; the adjacency list is slower to support this operation.

Adjacency Matrix

In mathematics and computer science, an adjacency matrix is a means of representing which vertices (or nodes) of a graph are adjacent to which other vertices. Another matrix representation for a graph is the incidence matrix.

Specifically, the adjacency matrix of a finite graph G on n vertices is the $n \times n$ matrix where the non-diagonal entry a_{ij} is the number of edges from vertex i to vertex j, and the diagonal entry a_{ii}, depending on the convention, is either once or twice the number of edges (loops) from vertex i to itself. Undirected graphs often use the latter convention of counting loops twice, whereas directed graphs typically use the former convention. There exists a unique adjacency matrix for each isomorphism class of graphs (up to permuting rows and columns), and it is not the adjacency matrix of any other isomorphism class of graphs. In the special case of a finite simple graph, the adjacency matrix is a (0,1)-matrix with zeros on its diagonal. If the graph is undirected, the adjacency matrix is symmetric.

The relationship between a graph and the eigenvalues and eigenvectors of its adjacency matrix is studied in spectral graph theory.

Adjacency Matrix of a Bipartite Graph

The adjacency matrix A of a bipartite graph whose parts have r and s vertices has the form

$$A = \begin{pmatrix} O & B \\ B^T & O \end{pmatrix},$$

where B is an $r \times s$ matrix and O is an all-zero matrix. Clearly, the matrix B uniquely represents the bipartite graphs. It is sometimes called the biadjacency matrix. Formally, let $G = (U, V, E)$ be a bipartite graph with parts $U = u_1,...,u_r$ and $V = v_1,...,v_s$. The biadjacency matrix is the r x s 0-1 matrix B in which $b_{i,j} = 1$ iff $(u_i, v_j) \in E$.

If G is a bipartite multigraph or weighted graph then the elements $b_{i,j}$ are taken to be the number of edges between the vertices or the weight of the edge (u_i, v_j), respectively.

Properties

The adjacency matrix of an undirected simple graph is symmetric, and therefore has a complete set of real eigenvalues and an orthogonal eigenvector basis. The set of eigenvalues of a graph is the spectrum of the graph. Suppose two directed or undirected graphs G_1 and G_2 with adjacency matrices A_1 and A_2 are given. G_1 and G_2 are isomorphic if and only if there exists a permutation matrix Psuch that

$$PA_1P^{-1} = A_2.$$

In particular, A_1 and A_2 are similar and therefore have the same minimal polynomial, characteristic polynomial, eigenvalues, determinant and trace. These can therefore serve as isomorphism invariants of graphs. However, two graphs may possess the same set of eigenvalues but not be isomorphic.

If A is the adjacency matrix of the directed or undirected graph G, then the matrix A^n (i.e., the matrix product of n copies of A) has an interesting interpretation: the entry in row i and column j gives the number of (directed or undirected) walks of length n from vertex i to vertex j. This implies, for example, that the number of triangles in an undirected graph G is exactly the trace of A^3 divided by 6.

The main diagonal of every adjacency matrix corresponding to a graph without loops has all zero entries. Note that here 'loops' means, for example A→A, not 'cycles' such as A→B→A. For (d)-regular graphs, d is also an eigenvalue of A for the vector $v=(1,\ldots,1)$, and $G$is connected if and only if the multiplicity of d is 1. It can be shown that $-d$ is also an eigenvalue of A if G is a connected bipartite graph. The above are results of Perron–Frobenius theorem.

Variations

An (a, b, c)-adjacency matrix A of a simple graph has $A_{ij} = a$ if ij is an edge, b if it is not, and c on the diagonal. The Seidel adjacency matrix is a (–1,1,0)-adjacency matrix. This matrix is used in studying strongly regular graphs and two-graphs.

The distance matrix has in position (i,j) the distance between vertices v_i and v_j. The distance is the length of a shortest path connecting the vertices. Unless lengths of edges are explicitly provided, the length of a path is the number of edges in it. The distance matrix resembles a high power of the adjacency matrix, but instead of telling only whether or not two vertices are connected (i.e., the connection matrix, which contains boolean values), it gives the exact distance between them.

Bibliography

Andrew Simpson: *Discrete Mathematics by Example*, McGraw-Hill Incorporated, New York, 2022.

Biggs, N.; Lloyd, E.; Wilson, R.: *Graph Theory,* Oxford University Press, Oxford, 2006.

Bojana Obrenic: *Practice Problems in Discrete Mathematics*, Prentice Hall, Nyew York, 2013.

Brian Hopkins: *Resources for Teaching Discrete Mathematics*, Mathematical Association of America, 2018.

Chartrand, Gary: *Introductory Graph Theory*, Dover, New York, 2005.

Crisler, N., Fisher, P., & Froelich, G.: *Discrete Mathematics through Applications*, W. H. Freeman and Company, New York, 2014.

Donald E. Knuth: *The Art of Computer Programming,* Addison-Wesley Professional, New York, 2019.

Dossey: Dossey, J., Otto, A., Spence, E., & Vanden Eynden, C.: *Discrete Mathematics*, Harper Collins College, New York, 2013.

Ebbinghaus, H.-D.; Flum, J.; Thomas, W.: *Mathematical Logic,* Springer, New York, 2014.

Enderton, Herbert: *A mathematical Introduction to Logic*, Academic Press, Boston, MA, 2011.

Gibbons, Alan: *Algorithmic Graph Theory*, Cambridge University Press, Cambridge , 2005.

Graham, Ronald L.; Groetschel, Martin; and Lovász, Lászlo: *Handbook of Combinatorics*, Elsevier, Amsterdam, NL, and Cambridge, MA, 2016.

Hamilton, A.G.: *Logic for Mathematicians,* Cambridge University Press, Cambridge, 2008.

Jioí Matoueek; Jaroslav Nesetoil: *Discrete Mathematics*, Oxford University Press, Oxford, 1998.

Kenneth H. Rosen: *Discrete Mathematics: And Its Applications*, McGraw-Hill College, New York, 2017.

Kenneth H. Rosen; John G. Michaels: *Hand Book of Discrete and Combinatorial Mathematics*, CRC Press, New York, 2014.

Kenney: Kenney, M. J., & Hirsch, C. R.: *Discrete Mathematics Across the Curriculum, K-12: 1991 Yearbook*, National Council of Teachers of Mathematics, Reston. VA, 2011.

Lindner, Charles C.; and Rodger, Christopher A.: *Design Theory*, CRC-Press; London, 2007.

Mark Newman: *Networks: An Introduction*, Oxford University Press, Oxford, 2020.

Martin Davis, Ron Sigal, Elaine J. Weyuker: *Computability, Complexity, and Languages: Fundamentals of Theoretical Computer Science*, Academic Press, New York, 2014.

McEliece: McEliece, R., Ash, R., & Ash, C.: *Introduction to Discrete Mathematics*, Random House, New York, 2009.

Mendelson, Elliott: *Introduction to Mathematical Logic,* Chapman & Hall, London 2017.

Norman L. Biggs: *Discrete Mathematics*, Oxford University Press, London, 2022.

Ralph P. Grimaldi: *Discrete and Combinatorial Mathematics: An Applied Introduction*, Addison Wesley, New York, 2014.

Riordan, John: *An Introduction to Combinatorial Analysis*, Wiley & Sons, New York, 1998.

Ronald Graham, Donald E. Knuth, Oren Patashnik, *Concrete Mathematics*, Oxford University Press, London,, 2017.

Samuel R. Buss: *Handbook of Proof Theory*. Elsevier, 2008.

Stanley, Richard P.: *Enumerative Combinatorics*, Cambridge University Press, Cambridge, 2017.

Susanna S. Epp: *Discrete Mathematics With Applications*, Thomson Brooks/Cole, London, 2020.

Troelstra A.S.; H. Schwichtenberg: *Basic Proof Theory*, Cambridge University Press, NY, 2016.

Tucker: Tucker, A.: *Applied Combinatorics*, John Wiley & Sons, New York, 2004.

Walicki, Micha: *Introduction to Mathematical Logic*, World Scientific Publishing, Singapore, 2018.

Wilson, Robin:*Four Colors Suffice*, Penguin Books, London, 2012.

Index

G

H

I

L

M

N

O

P

S

T

V

W

❑❑❑